Systemic Structural Constellations and Sustainability in Academia

In order to create truly sustainable universities, we require new methods of visualising and interpreting them holistically as institutions built on complex relationships and systems, rather than as individual departments and people operating independently. This book uses a systemic structural constellations approach to demonstrate how we can build more sustainable higher education institutions, both in terms of teaching and research and at an operational level.

Drawing examples from current research and teaching, *Systemic Structural Constellations and Sustainability in Academia* explores how universities are not only centres of teaching and learning but can also play a crucial role in enabling future decision-makers to appreciate and contribute to a more sustainable future.

Providing a clear introduction to systemic structural constellations and guidance on how to practically apply the theory to numerous aspects of the higher education system, this book will be of great interest to students and researchers of education for sustainable development, organisational learning and sustainable management, as well as those tasked with transforming the higher education system for the future.

Marlen Gabriele Arnold is a researcher and course co-ordinator at the Technical University Chemnitz, Germany.

T0347438

Systemic Structural Constellations and Sustainability in Academia

A New Method for Sustainable Higher Education

Marlen Gabriele Arnold

LONDON AND NEW YORK

First published 2017 by Routledge

2 Park Square, Milton Park, Abingdon, Oxfordshire OX14 4RN
52 Vanderbilt Avenue, New York, NY 10017

Routledge is an imprint of the Taylor & Francis Group, an informa business

First issued in paperback 2019

British Library Cataloguing in Publication Data
A catalogue record for this book is available from the British Library

Library of Congress Cataloging in Publication Data
A catalog record for this book has been requested

ISBN: 978-1-138-22394-3 (hbk)
ISBN: 978-0-367-27120-6 (pbk)

Typeset in Times New Roman
by Apex CoVantage, LLC

Contents

Introduction

Sustainability is neither anchored within academic systems nor comprehensively implemented in science and in teaching in Germany. In combining sustainability with the academic world, one point becomes very clear: Humans have gained so much knowledge about the world and cause–effect relations but are not able or willing to transfer it into practice and action accordingly and comprehensively. This limitation is not only caused by existing structures but influenced by previous or systemic factors. Thus, new methodologies and tools should be integrated into academia in order to foster sustainability and address a whole-institution approach. Even higher education institutions act more and more in multicausal and inter- and transdisciplinary contexts (Schneidewind et al. 2016) that require new and innovative methods. The integration of systemic structural constellations in research and teaching, as well as in administration, for initiating change processes allows new understanding, interventions and effective governance. How can systemic structural constellations foster change in academic systems towards sustainability? Academia can integrate sustainability challenges within two areas (Müller-Christ, 2014): (1) diverse contributions to a sustainable development (like science, education and higher education, transfer and consulting) and (2) sustainability of the academic institution and system itself (social, environmental and economic impact, organisational change and monitoring). The whole-institution approach "explicitly links research, educational, operational and outreach activities and engages students in each" (McMillen and Dyball, 2009: 55). Changing academic systems is always assigned a whole-institution approach and orientated on participation. Different stakeholder groups can work with this method as chairpersons, sustainability coordinators or sustainability change agents, student groups, researchers or administrative departments, and so forth. Systemic structural constellations provide a new approach, making reflexive and systemic elements tangible and creating transformational processes actively.

Using systemic constellations the pattern of relationships, structures, interaction, implicit knowledge and hidden or underlying dynamics and influences within a system can be made obvious by a way of representing. It is also a way of focusing masses of information and data, details and opinions and of pointing out aspects within a new dimension. In general, it is crucial to distinguish between different levels in institutional contexts having an influence on the current situation or problem, as well as on the solution. As in academic contexts, people are part of the main transaction and exchange processes as there is always a personal level. Yet, besides the personal issues, there is an organisational level that constellators should always have in mind, as well as the more complex system level (compared to family systems), because there are many more moderating factors and interrelations. In systemic structural constellations, systems can be simulated by the spatial arrangements of persons or symbols. The success of the method is related to action research and can be described in terms of the systematic spatial locations and perception of decision makers. Systems constellations permit both a deep look into the informal structures and relationships of institutions and social structures, as well as the testing of interventions or different solution options with regard to their effects.

When fostering sustainability, it should always be clear that sustainability is one topic among many others (like quality, development, finance, etc.) and has to be integrated on the basis of systemic principles within an institution. In complex contexts, such as sustainability and resilience, learning progress can be achieved on the knowledge level as well as on the action level. So integrating the new method of structural constellations can offer a fast growing problem identification and structuring. It further allows direct problem solving in cooperation with administration and science. The method appears to be a very powerful tool in explaining and transferring multicausalities in systems and offers itself as a complement to traditional methods. Findings, implementation options, conclusions and so forth resulting from work with systemic constellations are often not attainable by a pure study of documents, interviews or an empirical survey, at least not at a comparable speed. The visualisation of depth structures and hidden patterns allows a discussion of ways in which transition academia may move towards more sustainability. The use of the method requires sound training in systemic structural constellations and a respectful interaction with people and their concerns. Because the method has the potential to build a bridge between administration, science, innovation and change for identifying and initiating new options, as well as ways for action and implementation towards sustainability in academia, an ethical and respectful treatment of concerns, people and processes is mandatory. Moreover, system constellations do not automatically cause actions or changes towards more sustainability, but they enable new perspectives and behaviour. The efficacy goes beyond academia and thus stresses whole-institution approaches. Finally, the method is based on the voluntariness and willingness to participate. Moreover, profound training is necessary in order to conduct constellations and realise its guiding principles. It might also help to overcome fixed mental models, as well as resistance to change (Lozano et al., 2013), as it works with the unconscious and not against it.

The first part of the book is a reflection of sustainability challenges in academia, already reached milestones and necessary transition. It includes a theoretical reflection of constellations in academia, possible contributions concerning sustainability and their specific applications. Systemic structural constellations as a new method in academia will be introduced in Chapter 2. For interested readers, the next section discusses current ideas on how representative perception is working. It is followed by a methodical reflection on systems constellations and a discussion on where this new method fits in the already existing frameworks and classifications. Chapter 5 summarises essential systemic principles and highlights systems constellations in organisational or institutional contexts by discussing differences between family systems and business-oriented systems. Practical examples using constellations in sustainable academic contexts are presented in Chapter 6. Chapter 7 discusses the limits and possibilities of the new method based on whole-institution approaches, and in Chapter 8 several highlights are given and essential insights are summarised.

References

Lozano, R., Lukman, R., Lozano, F.J., Huisingh, D. and Lambrechts, W., 2013. 'Declarations for sustainability in higher education: becoming better leaders, through addressing the university system'. *J. Clean. Prod.* 48, 10–19.

Mc Millin, J. and Dyball, R., 2009. 'Developing a whole-of-university approach to educating for sustainability linking curriculum, research and sustainable campus operations'. *J. Educ. Sustain. Dev.* 3(1), 55–64.

Müller-Christ, G., 2014. *Nachhaltiges Management: Einführung in Ressourcenorientierung und widersprüchliche Managementrationalitäten.* UTB: Baden-Baden.

Schneidewind, U., Singer-Brodowski, M., Augenstein, K. and Stelzer, F., 2016. Pledge for a Transformative Science, 191_Wuppertal Paper, http://wupperinst. org/a/wi/a/s/ad/3554/.

1 Setting the scene
Sustainability challenges in academia

Sustainability challenges

Recent decades have shown that science is limited in prediction, planning and pinpoint governance. This becomes very obvious in the light of sustainability. Humans have gained considerable knowledge about the world and cause–effect relations but are not able or willing to transfer it into common practice and action. This limitation is not only caused by existing structures but influenced by past or systemic factors and has several reasons: the complexity and manifold understanding of sustainability or a sustainable development (Barth, 2015); the ongoing separation of logic and intuition (Rosselet, 2013), the focus on mental explanations and solutions rather on emotional or unconscious ones (see also Scharmer and Kaufer, 2013); a lack of understanding of systemic patterns and hidden agendas; organisational or institutional inertia or simply not enough money, and so on. Nevertheless, conventional scientific approaches and practices appear to be limited as they often do not really and comprehensively explain reality, ongoing behaviour, strategic decisions or the resilience of a system. Resilience includes not only (a) the extent of change or transformation while preserving the system structure and performance but also (b) the extent of the self-organisation of a system without regulating intervention (internal or external) and especially (c) the extent of learning and adaptability, the willingness to experiment and to implement new solutions (Walker and Carpenter, 2002). Flexible learning processes are essential for survival and vital for systems when facing new challenges. However, resilience is not necessarily a desirable state, as even system configurations can be highly resilient, a characteristic that is harmful to the common good (e.g. path dependence; Raven, 2007). Comparable problems emerge in the context of sustainability.

Neither sustainability nor sustainable development is clearly and unambiguously defined but is often part of the learning and negotiation processes (Stoltenberg and Burandt, 2014; Arnold and Barth, 2009). There are manifold understandings and definitions of sustainability, which are often aimed at the integration of environmental, social and economic concerns (Arnold, 2015). Among the principles of sustainable development are shaping, managing, producing and living within human systems in such a way that the ecological and social limits of carrying capacity are not exceeded (Allianz Sustainable Universities in Austria, 2014: 6). The Earth's ecosystems must be unharmed in their assimilation, buffering and regenerative capacity. The configuration of socially and economically more resilient systems is linked to this too. Accordingly, recognising, understanding, analysing, evaluating and creating sustainable contexts are of pivotal importance. As the main discussion is about economic, social and ecological capacity, we should think of using the term capacity or sustainable capacity in order to describe or better replace sustainability. Capacity is more aligned with the limits of all system levels, negotiations, values and the ongoing discourse about the substitution of different forms of capital in the context of a sustainable development. Sustainability is more aligned with maintenance – and this is a false precondition in an evolutionary world. However, as I do not want to disarray the context, I continue writing about sustainability, sustainable development and education for sustainable development.

Barth (2015) highlights that sustainability is always a transdisciplinary concept addressing three key issues: Sustainability science is (1) normative, (2) integrative and (3) participative. Sustainability science is based on human, individual and societal norms and values, as well as their reflection, and addresses normative positions for problem solving. Given complex interactions and dynamics, sustainability science aims at linking different forms of knowledge, levels of systems, disciplines and stakeholders in order to find integrative and stable solutions. Finally, a sustainable development is based on a whole-society approach where different interests, views and demands are discussed, balanced and realised in a participative way supporting societal learning. Therefore, according to Barth (2015), both learning and education play an important role. However, new methodologies and tools should be integrated into academia in order to foster sustainability, as higher education institutions act more and more in multicausal and inter- and transdisciplinary contexts that require new and innovative methods.

Education for sustainable development

At the Stockholm Conference in 1972, education for sustainable development was recognised as a vital topic for shaping the future (United Nations Environmental Program [UNEP], 1972). In 1992 in Agenda 21, the importance of education as a precondition for facilitating a sustainable development was formulated explicitly (Stoltenberg and Burandt, 2014: 567). In 2005, the topic of education for sustainable development once more became the focus of public attention by the United Nations. Implementing sustainability issues in learning and education was the main aim during the United Nation Decade of Education for Sustainable Development, 2005 until 2014. Four major thrusts and seven strategies were formulated (UNESCO, 2005):

Thrusts (UNESCO, 2005: 5)

- improving access and retention in quality basic education;
- reorientation of existing educational programmes to address sustainability;
- increasing public understanding and awareness of sustainability;
- providing training to advance sustainability across all sectors.

Strategies (UNESCO, 2005: 17)

- vision building and advocacy;
- consultation and ownership;
- partnership and networks;
- capacity building and training;
- research and innovation;
- use of information and communication technology (ICT);
- monitoring and evaluation.

Based on this variety, significant impulses were set during the decade, and at the same time different foci emerged at different levels and areas of education in the respective countries (UNESCO, 2014; Rieß, 2010; Schrenk and Holle-Giese, 2005). Commonly, there were manifold activities and initiatives, but a comprehensive and holistic concept is still missing (Grindsted and Holm, 2012; Scott et al., 2012).

The expansion of sustainability-related values, knowledge and skills belongs to the foundations of an education for sustainable development (Michelsen and Rieckmann, 2014). In particular, *Gestaltungskompetenz* (the ability to shape the future) is highlighted by UNESCO, for example skills such as forward thinking or the capacity for teamwork or enabling people to assess future decisions. Vare and Scott (2007) and Wals (2011) distinguish two current streams of education for sustainable development:

- One thread refers to the promotion of behavioural changes, as well as a behaviour pattern and mindset emerging within or based on social consensus processes. In this understanding, education for sustainable development can be seen as a learning for sustainable development.
- In the second thread, the development of skills is in the foreground, skills like critical thinking, reflecting and questioning concepts for sustainable development. It's all about discovering and handling dilemma. Learning as a sustainable development is central.

In particular, UNESCO (2014) made clear that a sustainable development and education for all are based on a special quality of education highlighting and communicating both knowledge and fundamental perspectives, as well as attitudes for a sustainable behaviour in all contexts (also see Mochizuki and Yarime, 2016; Ramos et al., 2015). Wals (2010) sees a new understanding of education, which gives room for discourse, debate and reflection. Since the understanding of education can vary depending on world views, the ideas of man, or humanity, as well as values, Stoltenberg and Burandt (2014) make clear that there is an understanding of the need to deal with these topics in a self-determined way in Western civilisation. That is why academia has such a pivotal importance in the context of a sustainable development. Yet clear, comprehensive strategies and a wide implementation within academia are very rare. Despite cultural and educational diversity, the international debate on education for sustainable development focuses on the following aspects (Stoltenberg and Burandt, 2014):

- *Focus on values*: The framework of values includes content, such as human dignity, resilience, ecological carrying capacity, distribution, access, diversity, like cultural diversity, and the like.
- *Educational objectives*: Education for sustainable development must enable citizens, learners and students to transfer the idea of a sustainable development actively in their contexts. This point is about the ability to successfully handle complex and interdisciplinary issues and to participate actively in social transformation processes.
- *Emphasis on competence*: In 2005, the Organisation for Economic Co-operation and Development (OECD) initiated a global discourse on educational requirements and strengthened the discussion on competencies. Competencies in terms of knowledge, skills, attitudes and

values and so-called key competencies create a foundation for sustainable development and social cohesion. Competence-oriented educational approaches focus on the desired result of educational processes, for example acquired competencies, and thus on the question of which problem-solving strategies, capacity to act and interpersonal competencies are important and should be fostered (de Haan, 2008).

- *Content of education*: Content of education and the provision of educational topics are not arbitrary but should: (1) be central to sustainable development processes, (2) have local and global references, (3) have long-term significance, (4) be interdisciplinary and (5) have the potential for action.
- *Learning processes*: Individual and social learning formal, non-formal and informal learning processes are desired.
- *Methods and methodologies*: The learning objectives are shifting from expertise to application-oriented contexts integrating a variety of forms of knowledge, such as system knowledge and the application of methods and procedures. This has an impact on the teaching and learning designs (use of methods and media) because competencies are less taught then acquired. Moreover, learning contexts should be established in which specific questions, problems or tasks are handled interdisciplinarily and are solution driven, minimising external effect and rebound effects.

According to Stoltenberg and Burandt (2014), education for sustainable development is thus neither an additional nor a new task for educational institutions but a deep change of perspectives stressing new thematic priorities or foci and methods or methodologies. Therefore, new media and their purposeful use for teaching sustainability-related contents and strengthening competencies are crucial (Möller, 2013). Education strategies for sustainable development should consequently address and integrate the aspects previously listed.

Education for sustainable development in academia

One of several goals in the decade of education for sustainable development is to anchor interdisciplinarity and innovative development processes in academia. Barth (2015: 19) highlights higher education in the context of education for sustainable development "since universities not only generate and transfer relevant knowledge, but in addition educate future decision-makers to enable them to contribute to a (more) sustainable future". In this context, the raising internationalisation and global knowledge transfer and exchange are also to be stressed. During the past decade, there were networking, teaching, science-based or project-based activities, and planning or strategic approaches were developed for implementing sustainability within higher education institutions. However, there is a huge variance from country to country (UNESCO, 2014, 2013). In particular, universities respond with innovative actions in teaching, such as project seminars, courses, ring courses or the use of new methods or working groups compiling sustainability concepts (Rowe and Hiser, 2016; Schneidewind et al., 2016; Holm et al., 2015; Ramos et al., 2015; Barth et al., 2014; UNESCO, 2014, 2013; Dlouhá et al., 2013; Lozano, 2006; Cortese, 2003). This diversity also shows the lack of coherent standards, indicators and strategies for education for sustainable development within academia. Thus, up to now, education for sustainable development found its way neither coherently nor holistically into the higher education landscape. According to Müller-Christ (2013), it implies for academia that the formulated willingness to integrate sustainability is clearly greater than the ability to act. Even the Higher Education Sustainability Initiative (HESI), established around the United Nations Conference on Sustainable Development (Rio+20) in 2012 in order to commit to the development of sustainable activities in academia, is not really fostered by the international academic community. Reasons for this are various since the integration of sustainability in academia needs different actors and various measures. Müller-Christ (2013) also points out that the competence orientation for creating a sustainable development is mainly required from policy and social institutions. Nevertheless, even without an overall political strategy to promote education for sustainable development at higher education institutions, academia has enough scope to focus on sustainability and to implement education for sustainable development in the respective academic spheres of activity. Building bridges between academia and society in light of sustainability is necessary for achieving these goals (Wiek et al., 2012; Wright, 2004).

Barth (2015: 46) emphasises three different levels for establishing sustainability in higher education institutions: (1) research on sustainability, (2) teaching and learning on sustainability topics and (3) institutional or organisational change processes "as self-reflective praxis, embracing management processes and operational parameters" of higher education institutions themselves, which are stakeholders in a sustainable society. Müller-Christ (2013) aggregates theses aspects and argues for two management approaches that can support the implementation of sustainability more deeply within academia (Müller-Christ, 2013): (1) diverse contributions to a sustainable development (like science, education and higher education, transfer and consulting) and (2) sustainability of the academic institution and system itself (social, environmental and economic impact, organisational change and monitoring).

- *Academia for a sustainable development*: As a social mission, sustainability gets more space within all academic tasks. The following questions can be guiding for change towards and implementation of sustainability issues: What kind of research contributes and leads to more sustainable development of society? What teaching and learning content should be included in all curricula? Which teaching and discussion methods are meaningful in the context of sustainability? How can funding and formation be supported?
- *Sustainable academia*: Specific measures to conserve resources or manage them sustainably at academia are addressed. Guiding questions can be as follows: How are the performance of academia and social responsibility secured sustainably? What is the social, ecological and economic impact of academia? How can the interdependency between anthroposphere and natural habitat be managed in a sustainable way?

Both perspectives require a sustainability management system for the following areas (Müller-Christ, 2013; also see Figure 1.1):

- formulation or adaptation of an academic mission statement highlighting sustainability;
- flow management of resources, energy and waste, input–output relations of environmental and social core areas;
- establishment of a systematic holistic and sustainability-oriented management system;
- open communication and sustainability reporting.

Sterling et al. (2016: 91) differentiate four broad categorisations of education for sustainable development based on research interests: (1) curriculum change and learning processes; (2) systemic change and institutional learning; (3) sustainability competence, action and engagement; and (4) institutional impact on the community and effecting change towards sustainability. So the authors set their focus on the more complex and emergent designed processes and institutional change processes.

While there are several concepts for analysing education in terms of sustainable development, integrated quality standards or coherent processes are still missing. Mochizuki and Yarime (2016) emphasise the

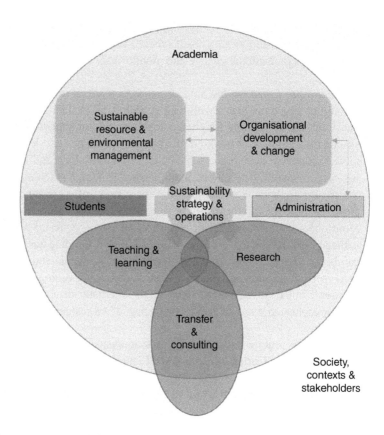

Figure 1.1 Academia in light of a whole-institution approach

science–policy gap and the need for transforming global governance in light of sustainability. Even a whole-institution approach "transforming the curricula, pedagogy, research and operations at the core of higher education and engaging all stakeholders – governing bodies, students, faculty and community" (UNESCO, 2014: 126) is rarely recognisable but highly appreciated. For instance, critical success factors for curriculum transformation are, among others, providing active participation and involvement, as well as creating opportunities for debating and discussing issues and recognising, reflecting and resolving concerns (de la Harpe and Thomas, 2009; de la Harpe and Radloff, 2003).

UNESCO (2014: 10) stresses that pedagogical innovation, like whole-institution approaches, has galvanised practices in education for sustainable development, as well as the use of interactive, learner-driven methods. "Student participation will be essential in scaling up whole-institution approaches, as these are proving to be highly effective methods to make sustainability issues a part of students' everyday experience. Research, documentation and sharing of experience will be essential to strengthen whole-institutional approaches" (UNESCO, 2014: 179). Finally, the transformation of policy, both institutional in general and educational in particular, is necessary (Mader et al., 2013; Lukman et al., 2010). Figure 1.1 stresses the need for a holistic academia-wide strategy integrating the institutional sphere (management, change and administration) with the duties and responsibilities of academia involving the main stakeholders by means of sustainable strategies and operations.

Changing academic systems is always assigned to a whole-institution approach and oriented on participation (Mc Millin and Dyball, 2009; D'Andrea and Gosling, 2005). D'Andrea stresses that a whole-institution approach "allows for different types and levels of devolution" and is not a one-size-fits-all model of changing academia but is based on interconnectedness and strategies for creating learning institutions (D'Andrea and Gosling, 2005: 6). Different stakeholder groups can work with this method as chairpersons, sustainability coordinators or sustainability change agents, student groups or researchers or administrative departments, and in other roles (Lozano et al., 2013; Stables and Scott, 2002). It is a way to incorporate sustainability into an extensive learning environment. In academia, the implementation of a whole-institution approach "will require deeper innovation in leadership and staff development" (UNESCO, 2014: 126), as well as peer learning across institutions. The UNESCO study also stresses that the whole-institution approach

> can best be achieved when multiple actors engage in a whole-system redesign. Such a redesign not only requires visionary leadership, social networking, new forms of research and high levels of participation, but also the introduction and support of interactive, integrative and critical forms of learning, in which multi-stakeholder social learning is exemplary.
>
> (UNESCO, 2014: 171)

Systemic structural constellations as a new method of addressing a whole-institution approach

According to Barth (2015), sustainability is always aligned with (1) complex and systemic problems, (2) interdisciplinarity, (3) self-directed and lifelong learning and (4) competence development. Sustainability challenges cannot be handled by monocausal thinking but by stressing uncertainty (Burandt and Barth, 2010), unforeseen dynamics as well as multilevel effects. Thus, people should be trained in recognising, understanding as well as dealing with these complex and systemic problems and possible solutions. This often requires interdisciplinary knowledge and negotiations (Jones et al., 2010). Education for sustainable development has to foster interdisciplinary collaborations too, that also take different cultures and mindsets into consideration. This goes in line with competence development emphasising the development of the whole personality in general and the abilities to identify, analyse, balance and manage, and otherwise deal with sustainability challenges in an adequate way (see *ACPA* Sustainability Learning Outcomes[1] or OECD Key Competencies,[2] or see Barth, 2015: 64). This results, in turn, in lifelong learning and knowledge acquisition (Blewitt, 2004). Therefore, self-directed learning strategies, as well as strategies on how to learn to learn, are crucial (Straka, 2000). Twenty years ago, Sterling (1996: 36) had already emphasised, "How people, institutions and communities interact – the hidden and operational curriculum – is all important and should engender a sustainability ethos that is lived and critically reflected upon." All in all, new methods are necessary to address all these different issues.

Systemic structural constellations are a new approach for integrating reflexive and systemic elements, as well as for actively shaping transformation processes, so that education for sustainable development can be seen and experienced (Arnold, 2016; Müller-Christ and Liebscher, 2015). This method enables us to work on and to reflect systemic linkages based on specific questions (Varga von Kibéd and Sparrer, 2014). Using the method of systemic structural constellations, the foci of a system can be especially represented and simulated through spatial arrangements or the physical layout of persons or symbols (Varga von Kibéd and Sparrer, 2014). Müller-Christ (2013) describes the method's performance by placing decision-makers to the gallery systematically and spatially. He furthermore stresses two types of new information created in a simple way: Systemic constellations allow both a deep look into the informal structures and relationships of institutions or social structures and the testing of interventions or various solutions with regard to their effects. Systemic constellations can also be utilised, specifically in the context of education for sustainable development, demonstrating effective approaches for overcoming resistance and addressing specific educational contents interdisciplinarily, reflexively and experientially too. Consequently, content- and structure-based recommendations for implementation in everyday university life can be developed. The success of the method is assigned to action research (Schlötter, 2005) and can be described by the systematic spatial locations and perception of decision makers (Varga von Kibéd and Sparrer, 2014; Müller-Christ, 2013).

The integration of systemic structural constellations in research and teaching, as well as in administration for initiating change processes, allows new understanding, interventions and effective governance. How can systemic structural constellations foster change in academic systems towards sustainability? Using systemic constellation, the pattern of relationships, structures, interaction, implicit knowledge and hidden or underlying dynamics and influences within a system can be made obvious by a way of representing. It is also a way of focusing a mass of information and data, details and opinions and pointing aspects within a new dimension. In general, it is crucial to distinguish different levels in institutional contexts having an influence on the current situation or problem as well as on the solution. As in academic contexts, where people are part of the main transaction and exchange processes, there is always a personal level. Yet, beside the personal issues, there is an organisational level that constellators should always have in mind, as well as the more complex system level (compared to family systems), because there are many more moderating factors and interrelations. In systemic structural constellations, systems can be simulated by spatial arrangements of persons or symbols. Now let's see how it works.

Notes

1 ACPA Sustainability Task Force: Student Learning Outcomes Assessment Materials Guidebook, www.myacpa.org/sites/default/files/ACPA_Sustainability.pdf.
2 The Definition and Selection of Key Competencies: Executive Summary, www.oecd.org/pisa/35070367.pdf.

References

Allianz Sustainable Universities in Austria, 2014. *Handbuch zur Erstellung von Nachhaltigkeitskonzepten für Universitäten, erstellt von der Arbeitsgruppe "Nachhaltigkeitskonzepte" der Allianz Nachhaltige Universitäten in Österreich*

Koordination: H. Kromp-Kolb, T. Lindenthal, L. Bohunovsky (BOKU) T. Weiger (Universität Salzburg).

Arnold, M., 2015. 'Fostering sustainability by linking co-creation and relationship management concepts'. *J. Clean. Prod.*, forthcoming (available online), doi:10.1016/j.jclepro.2015.03.059

Arnold, M., 2016. 'Systemische Strukturaufstellungen als neue Methode zur Förderung einer nachhaltigen Entwicklung'. *Ökologisches Wirtschaften*, forthcoming.

Arnold, M. and Barth, V., 2009. 'Klima-und umweltbezogene Lernprozesse in partizipativen Produktentwicklungsverfahren: Möglichkeiten und Grenzen'. *J. Soc. Sci. Edu.* 8(3), 80–92.

Barth, M., 2015. *Implementing Sustainability in Higher Education: Learning in an Age of Transformation.* Routledge: New York.

Barth, M., Adomßent, M., Fischer, D., Richter, S. and Rieckmann, M., 2014. 'Learning to change universities from within: a service-learning perspective on promoting sustainable consumption in higher education'. *J. Clean. Prod.* 62, 72–81.

Blewitt, J., 2004. 'Sustainability and lifelong learning'. In: J. Blewitt and C. Cullingford (eds.), *The Sustainability Curriculum: The Challenge for Higher Education.* Earthscan: London, 24–42.

Burandt, S. and Barth, M., 2010. 'Learning settings to face climate change.' *J. Clean. Prod.* 18(7), 659–665.

Cortese, A.D., 2003. 'The critical role of higher education in creating a sustainable future'. *Plan. High. Educ.* 31(3), 15–22.

D'Andrea, V. and Gosling, D., 2005. *Improving Teaching and Learning in Higher Education: A Whole Institution Approach.* Open University Press: Maidenhead, UK.

de Haan, G., 2008. 'Gestaltungskompetenz als Kompetenzkonzept der Bildung für nachhaltige entwicklung'. In: I. Bormann and G. de Haan (Hrsg.), *Kompetenzen der Bildung für nachhaltige Entwicklung: Operationalisierung, Messung, Rahmenbedingungen, Befunde.* VS Verlag für Sozialwissenschaften: Wiesbaden, 23–43.

de la Harpe, B. and Radloff, A., 2003. 'The challenges of integrating generic skills at two Australian universities'. *Staff and Ed. Dev. Intl.* 7(3), 235–244.

de la Harpe, B. and Thomas, I., 2009. 'Curriculum change in universities: why education for sustainable development is so tough'. *J. Educ. Sustain. Dev.* 3(1), 75–85.

Dlouhá, J., Huisingh, D. and Barton, A., 2013. 'Learning networks in higher education: universities in search of making effective regional impacts'. *J. Clean. Prod.* 49, 5–10.

Grindsted, T.S. and Holm, T., 2012. 'Thematic development of declarations on sustainability in higher education'. *Environ. Econ.* 3(1), 32.

Holm, T., Sammalisto, K., Grindsted, T.S. and Vuorisalo, T., 2015. 'Process framework for identifying sustainability aspects in university curricula and integrating education for sustainable development'. *J. Clean. Prod.* 106, 164–174.

Jones, P., Selby, D. and Sterling, S., 2010. 'More than the sum of their parts? Interdisciplinarity and sustainability'. In: P. Jones, D. Selby, and S. Sterling (eds.), *Sustainability Education: Perspectives and Practice across Higher Education.* Earthscan: London, Sterling, VA, 17–38.

Lozano, R., 2006. 'Incorporation and institutionalization of SD into universities: breaking through barriers to change'. *J. Clean. Prod.* 14(9), 787–796.

Lozano, R., Lukman, R., Lozano, F.J., Huisingh, D. and Lambrechts, W., 2013. 'Declarations for sustainability in higher education: becoming better leaders, through addressing the university system'. *J. Clean. Prod.* 48, 10–19.

Lukman, R., Krajnc, D. and Glavic, P., 2010. 'University ranking using research, educational and environmental indicators'. *J. Clean. Prod.* 18(7), 619–628.

Mader, C., Scott, G. and Dzulkifli Abdul Razak, D., 2013. 'Effective change management, governance and policy for sustainability transformation in higher education'. *Sustain. Account. Manag. Policy J.* 4(3), 264–284.

Mc Millin, J. and Dyball, R., 2009. 'Developing a whole-of-university approach to educating for sustainability linking curriculum, research and sustainable campus operations'. *J. Educ. Sustain. Dev.* 3(1), 55–64.

Michelsen, G. and Rieckmann, M., 2014. 'Kompetenzorientiertes Lehren und Lernen an Hochschulen – Veränderte Anforderungen und Bedingungen für Lehrende und Studierende'. In: F. Keuper and H. Arnold (Hrsg.), *Campus Transformation: Education, Qualification & Digitalization.* Logos Verlag: Berlin, 45–65.

Mochizuki, Y. and Yarime, M., 2016. 'Education for sustainable development and sustainability science: re-purposing higher education and research'. In: M. Barth, G. Michelsen, I. Thomas, and M. Rieckmann (eds.), *Routledge Handbook of Higher Education for Sustainable Development.* Routledge: London, 11–24.

Möller, A., 2013. 'Neue Medien in der Bildung für Nachhaltige Entwicklung'. In: N. Pütz, N. Logemann, and M.K.W. Schweer (Hrsg.), *Bildung für nachhaltige Entwicklung: Aktuelle theoretische Konzepte und Beispiele praktischer Umsetzung, Psychologie und Gesellschaft 11.* 223–238.

Müller-Christ, G., 2013. *Hochschulen und Nachhaltigkeit: Bremer Appell zum Ende der UN-Dekade "Bildung für nachhaltige Entwicklung",* www.wiwi.uni-bremen.de/gmc/pdf/HS_Bremer_Appell.pdf.

Müller-Christ, G. and Liebscher, A.K., 2015. 'Advanced training for sustainability change agents – insights and experiences from a seminar series using the method of systemic constellations'. In: W. Leal Filho et al. (Hrsg.), *Integrative Approaches to Sustainable Development at University Level: Making the Links.* Springer: Cham, Switzerland, 451–466.

Ramos, T.B., Caeiro, S., Van Hoof, B., Lozano, R., Huisingh, D. and Ceulemans, K., 2015. 'Experiences from the implementation of sustainable development in higher education institutions: environmental management for sustainable universities'. *J. Clean. Prod.* 106, 3–10.

Raven, R., 2007. 'Niche accumulation and hybridisation strategies in transition processes towards a sustainable energy system: an assessment of differences and pitfalls'. *Energy Policy* 35, 2390–2400.

Rieß, W., 2010. *Bildung für nachhaltige Entwicklung: Theoretische Analysen und empirische Studien. Internationale Hochschulschriften, Band 542.* Münster: Waxmann.

Rosselet, C., 2013. *Andersherum zur Lösung: Die Organisationsaufstellung als Verfahren der intuitiven Entscheidungsfindung.* Versus: Zürich.

Rowe, D. and Hiser, K., 2016. 'Higher education for sustainable development in the community and through partnerships'. In: M. Barth, G. Michelsen, I. Thomas, and M. Rieckmann (eds.), *Routledge Handbook of Higher Education for Sustainable Development.* Routledge: Oxford, 315–330.

Scharmer, O. and Kaufer, K., 2013. *Leading from the Emerging Future: From Ego-System to Eco-System Economics.* Berrett-Koehler Publishers: San Francisco.

Schlötter, P., 2005. *Vertraute Sprache und ihre Entdeckung: Systemaufstellungen sind kein Zufallsprodukt – der empirische Nachweis.* Carl-Auer-Verlag: Heidelberg.

Schneidewind, U., Singer-Brodowski, M., Augenstein, K. and Stelzer, F., 2016. Pledge for a Transformative Science, 191_Wuppertal Paper, http://wupperinst.org/a/wi/a/s/ad/3554/

Schrenk, M. and Holl-Giese, W. (Hrsg.), 2005. *Bildung für nachhaltige Entwicklung: Ergebnisse empirischer Untersuchungen.* Hamburg: Kovač.

Scott, G., Tilbury, D., Sharp, L. and Deane, E., 2012. *Turnaround Leadership for Sustainability in Higher Education.* Australian Office of Learning and Teaching: Sydney.

Stables, A. and Scott, W., 2002. 'The quest for holism in education for sustainable development'. *Environ. Educ. Res.* 8(1), 53–60.

Sterling, S., 1996. 'Education in change'. In: J. Huckle and S. Sterling (eds.), *Education for Sustainability.* Earthscan: London, 18–39.

Sterling, S., Warwick, P. and Wyness, L., 2016. 'Understanding approaches to ESD research on teaching and learning in higher education'. In: M. Barth, G. Michelsen, M. Rieckmann, and I. Thomas (eds.), *Routledge Handbook of Higher Education for Sustainable Development.* Routledge: Oxford, 89–99.

Stoltenberg, U. and Burandt, S., 2014. 'Bildung für eine nachhaltige Entwicklung'. In: H. Heinrichs and G. Michelsen (Hrsg.), *Nachhaltigkeitswissenschaften.* Springer: Berlin Heidelberg, 567–594.

Straka, G.A., 2000. *Conceptions of Self-Directed Learning: Theoretical and Conceptional Considerations.* Münster: Waxmann.

UNESCO, 2005. *United Nations Decade of Education for Sustainable Development (2005–2014): International Implementation Scheme.* UNESCO: Paris.

UNESCO, 2013 (Hrsg.). *Deutsche UNESCO-Kommission e.V., Hochschulen für eine Nachhaltige Entwicklung.* Bonn: VAS-Verlag.

UNESCO, 2014. *Shaping the Future We Want.* UN Decade of Education for Sustainable Development (2005–2014) Final Report, Paris, France.

United Nations Environmental Program, 1972. *Stockholm Declaration of the Stockholm Declaration on the Human Environment.* Author: Nairobi.

Vare, P. and Scott, W., 2007. 'Learning for a change: exploring the relationship between education and sustainable development'. *J. Educ. Sustain. Dev.* 1. Jg., Heft 2, 191–198.

Varga von Kibéd, M. and Sparrer, I., 2014. *Ganz im Gegenteil: Tetralemmaarbeit und andere Grundformen Systemischer Strukturaufstellungen – für Querdenker und solche, die es werden wollen.* Carl-Auer-Verlag: Heidelberg.

Walker, B., Carpenter, S., Anderies, J., Abel, N., Cumming, G., Janssen, M., Lebel, L., Norberg, J., Peterson, G. D. and Pritchard R., 2002. 'Resilience management in social-ecological systems: a working hypothesis for a participatory approach'. *Conserv. Ecol.* 6(1), 14.

Wals, A.E.J., 2010. *Message in a Bottle: Learning Our Way Out of Unsustainability.* Wageningen UR: Wageningen.

Wals, A.E.J., 2011. 'Learning our way to sustainability'. *J. Educ. Sustain. Dev.* 5(2), 177–186.

Wiek, A., Farioli, F., Fukushi, K. and Yarime, M., 2012. 'Sustainability science: bridging the gap between science and society'. *Sustain. Sci.* 7(1), 1–4.

Wright, T.S.A., 2004. 'The evolution of environmental sustainability declarations in higher education'. In: A.E.J. Wals and P.B. Corcoran (eds.), *Higher Education and the Challenge of Sustainability: Problematics, Promise, and Practice.* Kluwer Academic Publishers: Dordrecht, The Netherlands, 7–19.

2 An introduction to systemic structural constellations

The potential of constellations

According to Wade (2004: 194), structural constellations "provide powerful and creative ways to clarifying and resolving complex, possibly intractable issues associated with organisations", systems or social actors. By means of using systemic constellation, the pattern of relationships, structures, interaction, implicit knowledge and hidden or underlying dynamics and influences within a system can made obvious by a way of representing. It is also a way of focusing a mass of information and data, details and opinions and pointing aspects within a new dimension. Wade (2004: 194) also highlights that, "apart from bringing clarity, constellations give opportunities to experiment with possible options in a safe environment to aid decision-making". Systemic constellations are able to represent and picture spatially patterns, relations, structures and relationships within a system, and thus they can be used for manifold issues (Kopp, 2013). Rosselet (2013: 16) recommends using systemic structural constellations whenever the following circumstances obtain:

- *Wealth of information*: There are several and diverse conceptualisation of problems and solutions.
- *High complexity*: There is a multitude of elements and relations that realise ongoing non-predictable patterns and structures.
- *Turbulences*: Uncertainty exists due to the instability of conditions and settings and/or the contingency of certain occurrences and events.

Thus, it seems this method is highly applicable for sustainability issues and whole-institution approaches.

Constellations work has mixed roots (Daimler, 2014), such as family systems therapy (e.g. psychodrama and group interaction) (Mainyu, 2011; Moreno, 1993; Satir, 1988; Boszormenyi-Nagy, 1987; Boszormenyi-Nagy and Spark, 1973), solution-focused concepts (Berg, 2006; de Shazer, 1994) and existential-phenomenology approaches (Husserl, 2003; Brentano, 1967; see Figure 2.1). Today there is a wide application of this method in different contexts like psychology, medicine, pedagogy, business management, and so on, yet structural constellations have some aspects in common. Structural constellations are also based on systems theory, highlighting the idea that systems are mainly self-regulating entities always seeking balance and ready to change or transform. Certain elements have a key role for balancing a system, so these items are system-relevant. According to Sterling (1996), education for sustainable development also relates to systems theory.

As there is a wide variety of structural constellations in practice (Daimler, 2014; Varga von Kibéd and Sparrer, 2014), the method is not arbitrary but has a certain frame (see Figure 2.1). In general, there are three different parts in structural constellations: a facilitator or process manager or constellator, an issue-holder and a group of people willing to participate as representatives in a constellation (see Figure 2.2). In science, it is also possible to work with students as representatives in teaching contexts. Systemic structural constellations are able to highlight research questions and test assumptions in order to understand a problem more deeply. Moreover, interventions can be tested, or new research propositions or hypothesises can be developed.

There are some phases to consider: First, the constellator and the issue-holder have a short conversation clarifying what the exact issue is, what kind of goal should be reached and what key elements should be involved in the constellation process. The next step is the selection of the people who should represent the core elements of the specific system. Selection is normally done by the issue-holder. Therefore, following her own intuition, the issue-holder asks individual persons to represent particular elements of the system. Anyone can reject the request, so that the issue-holder has to ask other persons to be part of the group. Finally, all representatives are selected and agree to join the constellation. (By the way, every person may ask to leave the constellation and may be replaced by another person.) Then the issue-holder places all the representatives in the room. That special grouping is called a structural constellation and is a spatial view that makes arrangements, distances, and directions and underlying dynamics of the situation more visible. In the case of personal

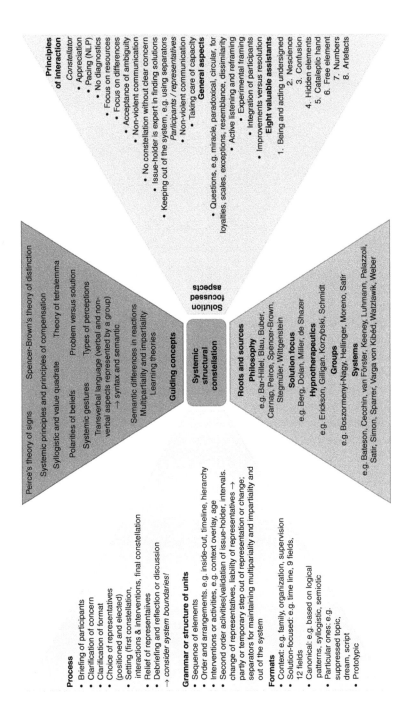

Figure 2.1 Overview of guiding aspects in light of systemic structural constellations

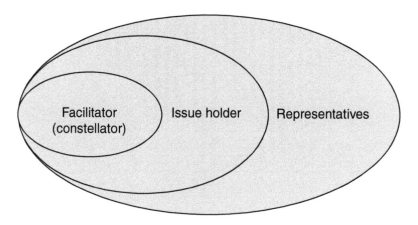

Figure 2.2 Involvements in a constellation

concerns, the issue-holder is also represented by another person (mainly the first time before the issue-holder steps into the constellation herself/ himself). Once placed, the representatives provide important data on the system by being resonators of implicit and hidden knowledge (Rosselet, 2013); this is also called representative perception (more about this in Chapter 3). The constellator either asks the representatives to follow their moving impulses or waits before doing so. Essential to the process is to allow representatives to describe the body perception, emotions or emerging images, sayings or sentences without any interpretation and evaluation. The use of non-violent communication should be a goal (Rosenberg, 2015). The representatives should say everything that makes a difference and occurs on the places where they stand – even if it does not make sense to them. Body signals (e.g. looking down at the ground, in the air, etc.) should be recognised attentively by the constellator as they might contain potential information for the ongoing process. After having a first view of the system, the ongoing processes and intervention differ from school to school and from constellator to constellator. The representatives can move according to their own impulses, and they can talk with one another during this search process; the constellator guides the process by suggesting changes in position or by recommending particular words and sentences. During the process, some representatives can be removed or added, or those representatives who were just selected but not con-stellated can be integrated within the process by being either asked or assigned a place. You can work with open elements or hide the meaning or representation of elements. In hidden representations, the constellator should at least pass a sheet of paper on which all meanings of representations are given to the audience or group of students or other stakeholders. Not having such a handout could act as a pall on the constellation process. The constellation process is usually completed when a solution or picture is found with which the issue-holder, or client, is satisfied. Or the process runs out. Systemic structural constellations take between 20 minutes and 2 hours. The representatives are released from their function, and the issue-holder, or client, thanks them. Finally, there is a debriefing between constellator and client (and joint researchers or participating people).

Moreover, different formats can be used for constellations (see Figure 2.1). Their choice strongly depends on the concern and the particular interest of the issue-holder as well as the constellator's familiarity with the format. Here, tetralemma, as one format of plenty of formats of constellations, will be introduced for showing particular elements. Tetralemma is always a good choice for decision making, the coalescence of differences or oppositions, the clarification of points of view, the check of values, the overcoming of blockades, the visualisation of non-observance, the integration of unlearnt issues, and the like (Daimler, 2014). Four different elements are distinguished and set as places: *this one*, *the other*, *both*, *neither* (see Figure 2.3). These four elements are set in a quadrat constituting the space, so that the focus (the representative of the issue-holder) can walk through in order to find solutions. The *fifth element* is a free element

Figure 2.3 Tetralemma as one format of constellations

(see Figure 2.3) representing *none of this and not even that* solution. It is free to walk around and do whatever the element wants to do, except moving the four placed elements. Within the constellation process, the focus can walk from element to element, relate with them accordingly and also as feedback. The representatives of the four places follow their impulses and articulate their representative perceptions. According to the textbook, the focus normally starts by *this one* followed by the *other*, continuing with *both* and finally *neither*, but actually the sequence can be freely chosen. The constellation comes to an end when the focus is satisfied with the interaction and has found a solution for her or him.

Aspects to highlight

There are some further and remarkable things to know:

- Everybody can contribute as a representative, but there are some obvious differences in expression or communication. As representation is mainly characterised by expressing emerging feelings, pictures, phrases, and the like, as strange as they might seem, that are not their own but belong to this specific context, some people do not try to express these emotions or changes with the same energy as when they are not representing another person or element but themselves. In this case, the facilitator or constellator has to be very attentive and can offer words, feelings or other options to support the process of expression.
- Moreover, research-oriented and solution-focused methods or learning contexts are useful and even more attractive to students because they can discover new patterns, relations, and so on or are able to reframe settings. In addition, academic members are able to discuss other or new relations and options that occurred during a constellation process.
- Furthermore, a constellation shows just a specific context resulting from the initial question or system focus. Observable interactions, structures, patterns or relationships cannot be transferred to other contexts, which means that, in a constellation, the focus is set on the question of how best to integrate sustainability into academia; people's behaviour, attitudes, expressions, feelings, and the like must not be adapted to questions around the implementation of, say, a quality system in academia. So, in constellation processes, it should always be emphasised that this specific context may not be easily transferred to other contexts or settings.

- Finally, representative perception has nothing in common with role-plays or theatre. There is nothing to learn; there are no given interactions or commands or determined settings (see also Chapter 3). It is just about the emerging feelings, pictures and moving impulses represented by persons that belong only in this specific context.

The following are eight helpful types of assistants (see Figure 2.1), each described briefly:

1 *Being and acting undersigned:* This attitude clarifies the role in the whole process and enables acceptance of non-guidance but attendance. It is about creating space for letting something happen and develop. It is not to be mixed up with being helpless and irresponsible during the process.
2 *Nescience:* This is a basic attitude as a constellator – not to know but to remain open and interested. It helps to prevent interpretations, prejudices, evaluations and hypotheses within the process. This attitude helps to accept things as they are. It supports new knowledge, astonishment and the relativity of awareness. The attitude of nescience is not to be mixed up with poor knowledge of techniques and instruments or bad education.
3 *Confusion:* Confusion is associated with lateral thinkers, the messenger of transformation and fruit of paradox and ambiguity (Varga von Kibéd, 2013). It demands to stay connected with the system and fosters deep change and learning as well as disengagement from dysfunctional control. Confusion is not to be mixed up with ignorance or lacking the capability to manage complexity.
4 *Hidden elements:* As previously stated, constellations with hidden elements are good practice in first-contact settings with students or people who are very sceptical. Moreover, they can be easily practiced in settings where nobody needs to know about the particular topic or when working with team members.
5 *Cataleptic hand:* The constellator uses her own hand for testing context overlays and separating contexts. Therefore, she brings one of the hands in a cataleptic state, eliminating the notion that the hand is still part of her own body. So it becomes an independent assistant. Then the constellator can put her cataleptic hand wherever necessary around an element (a representative) and draw something or a new element out of the representative. When the cataleptic hand is used further interventions depend on the intensity a representative reacts.

In case the representative follows the cataleptic hand, a new or additional element/ representative can be positioned within the constellation. Cataleptic hands are always useful for separating issues, like a team member is behaving more like the boss than the boss himself or is testing the relevance of new contexts, for example introducing a new element without choosing a new representative or when short of representatives.

6   *Free elements:* These are helpful because they are not bound to rules or process settings and can move independently. These elements just follow their own impulses and are allowed to do everything in an appreciative manner. Free elements can represent hyperpersonal vigour, like wisdom or freedom, or they can even work as a coach, emphasising crucial relations, differences, patterns, emotions and so on. They can also be used in early settings when a constellator is not that experienced.

7   *Numbers:* Questions for numbers can enhance clarity in stagnant or confusing situations and foster process towards a solution. Asking for numbers can be an essential hint for an alter ego context, as well as for the level of involvement of the representatives or a solution that was found (further, see Daimler, 2014; Erickson, 1938).

8   *Artefacts:* It is really helpful to support the process and to foster representation by using artefacts like puppets, different kind of weapons, crowns or simply stuff that we know from childhood. Having the opportunity to choose an artefact, the representatives can often express themselves better and get more clarity. This clarity is helpful for the ongoing process or simulations.

References

Berg, Insoo K., 2006. *Familien-Zusammenhalt(en): Ein kurztherapeutisches und lösungsorientiertes Arbeitsbuch.* Dortmund: modernes lernen.

Boszormenyi-Nagy, I., 1987. *Foundations of Contextual Therapy: Collected Papers of Ivan Boszormenyi-Nagy, MD.* Brunner/Mazel: New York.

Boszormenyi-Nagy, I. and Spark, G.M., 1973. *Invisible Loyalties: Reciprocity in Intergenerational Family Therapy.* Harper & Row: Hagerstown, MD.

Brentano, F., 1967. 'The true and the evident'. *Brit. J. Philos. Sci.* 18(3), 255–257.

Daimler, R., 2014. *Basics der Systemischen Strukturaufstellungen.* Kösel: München.

de Shazer, S., 1994. *Words Were Originally Magic.* Norton: New York.

Erickson, M.H., 1938. 'A study of clinical and experimental findings on hypnotic deafness: I. Clinical experimentation and findings'. *J. Gen. Psychol.* 19(1), 127–150.

Husserl, E., 2003 (first 1891). *Philosophy of Arithmetic.* Willard, Dallas, Dordrecht: Kluwer.

Kopp, U., 2013. 'Systemische Nachhaltigkeitskompetenzen für Führungskräft, Erfahrungen mit Aufstellungsarbeit in der Managementaus-und weiterbildung'. *Die Unternehmung* 67 Jg., Heft 2, 127–154.

Mainyu, E.A., 2011. *Jacob L. Moreno: Erikson's stages of psychosocial development, Teacher, Psychodrama, Group psychotherapy.* Aud Publishing.

Moreno, J.L., 1993. *Who Shall Survive? Foundations of Sociometry, Group Psychotherapy and Sociodrama.* American Society of Group Psychotherapy and Psychodrama: McLean, VA (first published in 1934).

Rosenberg, M.B., 2015. *Nonviolent Communication: A Language of Life (Nonviolent Communication Guides).* Puddle Dancer Press: Encinitas, CA.

Rosselet, C., 2013. *Andersherum zur Lösung: Die Organisationsaufstellung als Verfahren der intuitiven Entscheidungsfindung.* Versus: Zürich.

Satir, V., 1988. *The New Peoplemaking.* Science and Behavior Books: Palo Alto, CA.

Sterling, S., 1996. 'Education in change'. In: J. Huckle and S. Sterling(eds.), *Education for Sustainability.* Earthscan: London, 18–39.

Varga von Kibéd, M., 2013. *Semantische Reaktionen: Praxis des Systemischen Denkens I.* SySt®: München.

Varga von Kibéd, M. and Sparrer, I., 2014. *Ganz im Gegenteil: Tetralemmaarbeit und andere Grundformen Systemischer Strukturaufstellungen – für Querdenker und solche, die es werden wollen.* Carl-Auer-Verlag: Heidelberg.

Wade, H., 2004. 'Systemic working: the constellations approach'. *Ind. Commer. Train.* 36(5), 194–199.

3 The phenomenon of representative perception

Using the method of systemic structural constellations, implicit knowledge, as well as hidden or underlying but not consciously perceived dynamics and effects – comparable to the unknown region of the Johari window (Rosenberg, 2015) – can be depicted by so-called representative perception (Sparrer and Varga of Kibéd, 2010). Representative perception can be described by the body symptoms or feelings shown or verbalised by people who do not represent themselves but who are very consistent with the feelings of a person for whom they are representative in the system. People who do not know systemic structural constellations often react strangely, with incomprehension and doubt when told that in constellations people suddenly perceive feelings consistent with the feelings of other persons. This initial scepticism can often be reduced or abolished in the event that people come to personally experience the related outside perceptions and feelings. Using the method of systemic structural constellation can lead to both experiencing and representing a new dimension of human perception, as well as a new knowledge quality. This is comparable to the infrasound range (below the hearing limit of humans) or the ultrasound range (above the hearing limit), which is inaudible to humans but audible to other living beings – or that is produced by plants in extreme situations.

How representation works exactly is still under research and not understood as yet. A basic assumption of constellation work is that representative perception is not random or arbitrary, or even scripted, but determined by the position and the relations in the system itself. Schlötter (2005) proved this in an empirical test design meeting current scientific quality criteria. He demonstrated that the perception of the positions in a room is a kind of non-verbal language involving the positions of people in the room relative to one another or a kind of sign language or sign system, comparable to a language that follows generally understandable semantics. It should be stressed that representative perception works independently of language and culture. In his research project with 250 volunteers and more than 4000 individual tests, Schlötter (2005) was able to demonstrate that systemic constellations provide information that is not dependent on individual persons but mainly objective. There is no information or characteristics of individuals represented. However, there is no theory explaining the phenomenon of representative perception, as well as the very precise spatial reproduction of systems and appropriate developmental steps within the process work. The English biologist Ruppert Sheldrake (2001, 1988) coined the term morphic field, or morphic resonance, capable of retaining information. These fields are widely independent of space and time, enabling a kind of memory. However, it remains unclear where this field comes from, why it appears just in constellations, if and how it appears otherwise and what the nature of the information is.

Another current attempt to explain representative perception refers to the quantum field theory. Quantum entanglement, quantum tunnelling or quantum coherency is part of macro- as well as micro-systems and an essential research area in physics (Southwell, 2008; Vedral, 2008). "Quantum coherence is the idea of quantum entities multitasking. It's the quantum skier. It's an object that behaves like a wave, so that it doesn't just move in one direction or the other, but can follow multiple pathways at the same time" (Al-Khalili, 2015, 12:37 ff.). Gehlert (2014: 1) stresses that "Greenberg et al. (1990, 2008) published papers in which they demonstrated entanglement and quantum information transfer in multi-particle systems". He concludes further (2014: 5) that their

> experiments represent arbitrarily multipartite quantum fields. Now assuming that living organisms represent macroscopic quantum fields, due to their metastable, electromagnetic structure of their basis elements (atoms, molecules, synapses) and that they interact with each other, then the way systemic constellations work could be analogously put into relation. All currently observed phenomena, in the context of systemic constellations, could comprehensibly be explained and this, being on the basis of natural sciences.

That means that multi-particle systems are also quantum fields and information is spread and available immediately. In a quantum field, everything is connected, and everything can communicate with all. Gehlert (2014) argues further that just the observer interprets all information herself and that all interpretation is context related. Consequently, systemic structural constellations enable a metacommunication on social systems and provide a form of self-reflection on social systems supporting problem-solving contexts.

According to Matthias Varga von Kibéd (personal interview in autumn 2015), representative perception is something normal, natural, something that all people can do but nothing mystical or a mystical transfiguration. He argues that the understanding of perceptions should be more distinguished in Western and Eastern concepts as the focus is set more on the individual in the West, while in the East it is more focused on the group. Thus, it is an ability that we can use or that we are not used to or familiar with but able to use. Based on his work, Schlöter (2005: 5) assumes that constellations are based on immemorial archaic phenomena where at one time special relations between people took on an importance for social behaviour and were transformed into verbal metaphors.

However representative perception works, constellations were investigated as therapeutic-advisory in a set of actions at the University of Heidelberg (Weinhold et al., 2014). The primary results are (1) long-term improvement of mental state, (2) increased achievement of subjective goals and (3) an improved perception of interpersonal relationships. Thus, systemic constellations and representative perceptions seem to be very capable methods for reaching goals, stressing long-term satisfaction and relationships. Especially in academia, the achievement of goals, as well as the establishment of stable and constructive relationships, is of pivotal importance.

References

Al-Khalili, J., 2015. 'How quantum biology might explain life's biggest questions'. *TEDGlobalLondon.* 16:09. Filmed June 2015, www.ted.com/talks/jim_al_khalili_how_quantum_biology_might_explain_life_s_biggest_questions#t-308120.

Gehlert, T., 2014. 'GHZ–theorem and systemic constellations quantum teleportation in multi-particle systems without Bell's inequality', 10/2014; *ResearchGate*, doi:10.13140/2.1.4057.1848

Greenberger, D.M., Horne, M.A., Shimony, A. and Zeilinger, A., 1990. 'Bell's theorem without inequalities'. *Am. J. Phys.* 58, 1131–1143.

Greenberger, D.M., Horne, M. and Zeitiger, A., 2008. 'A Bell theorem without inequalities for two particles, using efficient detectors'. *Phys. Rev. A* 78.

Rosenberg, M.B., 2015. *Nonviolent Communication: A Language of Life (Nonviolent Communication Guides).* Puddle Dancer Press: Encinitas, CA.

Schlötter, P., 2005. *Vertraute Sprache und ihre Entdeckung: Systemaufstellungen sind kein Zufallsprodukt – der empirische Nachweis.* Carl-Auer-Verlag: Heidelberg.

Sheldrake, R., 1988. *The Presence of the Past.* Icon Books: London

Sheldrake, R., 2001. 'Das Morphologische Feld sozialer Systeme'. In: G. Weber (Hrsg.), *Derselbe Wind lässt viele Drachen steigen.* Carl-Auer-Verlag: Heidelberg, 29–42.

Southwell, K., 2008. 'Quantum coherence'. *Nature* 453(7198), 1003. doi:10.1038/4531003a

Sparrer, I. and Varga von Kibéd, M., 2010. *Klare Sicht im Blindflug – Schriften zur Systemischen Strukturaufstellung.* Springer: Heidelberg.

Vedral, V., 2008. 'Quantifying entanglement in macroscopic systems'. *Nature* 453(7198), 1004–1007. doi:10.1038/nature07124

Weinhold, J., Bornhäuser, A., Hunger, C. and Schweitzer, J., 2014. *Dreierlei Wirksamkeit, Die Heidelberger Studie zu Systemaufstellungen.* Springer: Heidelberg.

4 Methodical reflections on systemic constellations

Systemic structural constellations are part of qualitative research; they can be used in an exploratory and hypothesis-generating way. They can be assigned to field and action research (Bryman, 2012). Currently, qualitative research in socio-science and arts is ostensibly focused on conscious data (see also Christie and Miller, 2016; Kyburz-Graber, 2016; Rowe and Hiser, 2016; Wahr and de la Harpe, 2016). Up to now, there's often been double hermeneutics (Ginev, 2007; Giddens, 1984), as researchers make either interpretations or socially adapted evaluations of interviewees or involved persons in research settings. However, neuroscience has showed that our decisions were prepared by our unconsciousness (Soon et al., 2008).[1] Thus, while reducing one level of hermeneutics and finding stable or sustainable solutions, the unconsciousness should be taken more into account in research settings. Integrating unconscious interaction and decision making in scientific research enables more reliable and valid knowledge and can foster efficient and effective transformation. The strong focus on conscious knowledge, behaviour and decision making can be one reason for all the difficulties in implementing sustainability issues within a broad scope of action, as well as the remaining gap between explicit knowledge and behaviour. Thus, a stronger emphasis and embedment of unconscious knowledge and decision-making in research designs will be necessary.

So constellations can be used for organisational contexts having only limited information or current theories and concepts reaching their limits. Intuition and abduction can have an effect by uncovering hidden information via 3D images, as well as spatial language. Reichertz (1999: 47) argues against the alleged understanding that abduction would produce new knowledge using logical deduction. He worked out that abduction is a mental process, a spiritual act, a mental leap bringing together things that one never thought that belonged together (Reichertz, 1999: 54). He clarifies that a linguistic hypothesis is only the effect of an abduction and results from processes that are neither rationally justifiable nor criticisable. These kinds of processes are given in structural constellations as often new relations and combinations are visible, and thus new perspectives emerge. Furthermore, Sahm and von Weizsäcker (2016) showed that intuition plays an important role in the early and late stages of learning processes and has a crucial impact on decision making.

Constellations have a hypothesis-testing character to the effect that existing hypotheses can be tested to see whether the (often rational or consciously taken) assumption is confirmed by means of a constellation or very different contexts arise. In single settings, this form of hypothesis testing cannot meet scientific quality criteria in total. However, conducting constellations in the same format with the same question performed with various companies and/or company representatives belonging to the same industry qualifies the setting as meeting the scientific quality criteria. Systemic constellations should follow specific rules and guidelines, have to be managed without any interest in the solution by the constellator and should be in close feedback with the issue-holder(s). In this sense, it is a reliable method. In total, the method of systemic structural constellation has to follow scientific quality criteria in order to generate implicit or tacit knowledge in a stable way.

- Objectivity, a degree of quality representing how independent results are from the investigator and the basic conditions or boundary conditions, is differentiated concerning execution, analysis and interpretation (Bryman, 2012). According to Schlötter (2005), systemic structural constellations have a high objectivity. The practice of constellation is based on rules and detailed formats (Varga von Kibéd and Sparrer, 2014).
- Research reliability, the achievement of measurement results under the same conditions with identical measurement methods, is very sophisticated and complex in the context of constellations (Schlötter, 2005). In social contexts, reliability is characterised by a higher variance because, although constellations can be repeated with the same people, at the same time a learning or cognitive process is also happening, so the initial situation can never be reconstituted. Schlötter (2005) was able to create high reliability with his parallel tests and different test persons. In reliable settings, system constellations generate authentic information on implicit or unconscious knowledge, transforming it into explicit knowledge.

• Validity is a quality criterion for measuring the extent to which a measurement method actually detects the raised construct. Internal validity is closely connected with reliability (Lamnek, 1988). In highly complex situations – given the context of constellations – the well-known reliability–validity dilemma occurs. Concerning external validity, it should always be considered that the issue-holders assess the efficacy in and of the systemic constellation – in the sense, "Were the statements and the solution(s) helpful to make socioeconomic processes more resilient, sustainable and more effective?"

However, the balance of all quality criteria is crucial. A high validity is always to be secured by the issue-holder confirming the representation of the system. Having no external issue-holder or working with prototypic constellations, there should be a mindfulness of interpretation, since it is unclear what is represented in the field, and that Popper's (2002) stressing that scientists cannot encompass an objective is an absolutely valid truth should be kept mind. However, systemic constellations can strike a new path towards the novel in the understanding of science, as well as the centre of abduction and intuition again, beside deduction and induction. So even though deduction and induction are ongoing essential tools in scientific work, abduction, as well as intuition, needs to be placed more centrally in order to generate innovation, new strategies and new methods to foster sustainability comprehensively. This is in the context of Popper's (2002) emphasising the creative intuition of any discovery instead of the application of logical and rational methods, but it also stresses the necessity of developing new hypotheses and approaches by rules and guidelines as well as being rationally justified. Thus, there is a great potential in generating new sustainability-related knowledge and activities.

Note

1 Daimler (2014: 63 ff.) stresses leadership should always be connected to new insights of brain research.

References

Bryman, A., 2012. *Social Research Methods*. 4th ed. Oxford: Oxford University Press.
Christie, B. and Miller, K., 2016. 'Academics's opinions and practices of education for sustainable development. Reflections on a nation-wide, mixed-methods, multidisciplinary study'. In: M. Barth, G. Michelsen, M. Rieckmann, and

I. Thomas (eds.), *Routledge Handbook of Higher Education for Sustainable Development.* Routledge: Oxford, 396–410.

Daimler, R., 2014. *Basics der Systemischen Strukturaufstellungen.* Kösel: München.

Giddens, A., 1984. *The Constitution of Society: Outline of a Theory of Structuration.* The University of California Press: Berkeley.

Ginev, D., 2007. 'Doppelte Hermeneutik und Konstitutionstheorie'. *Deutsche Zeitschrift für Philosophie*, 55(5), 679–688.

Kyburz-Graber, R., 2016. 'Case study research on higher education for sustainable development: epistemological foundation and quality challenges'. In: M. Barth, G. Michelsen, M. Rieckmann, and I. Thomas (eds.), *Routledge Handbook of Higher Education for Sustainable Development.* Routledge: Oxford, 126–141.

Lamnek, S., 1988. *Qualitative Sozialforschung, Band 1: Methodologie.* BeltzPVU: München.

Popper, K., 2002. *The Logic of Scientific Discovery, Routledge Classics.* 2nd ed. Routledge: London.

Reichertz, J., 1999. 'Gültige Entdeckung des Neuen?: zur Bedeutung der Abduktion in der qualitativen Sozialforschung'. *Österreichische Zeitschrift für Soziologie* 24(4), 47–64. URN: http://nbn-resolving.de/urn:nbn:de:0168-ssoar-19536.

Rowe, D. and Hiser, K., 2016. 'Higher education for sustainable development in the community and through partnerships'. In: M. Barth, G. Michelsen, I. Thomas, and M. Rieckmann (eds.), *Routledge Handbook of Higher Education for Sustainable Development.* Routledge: Oxford, 315–330.

Sahm, M. and von Weizsäcker, R.K., 2016. 'Reason, intuition, and time'. *Manage. Decis. Econ.* 37(3), 195–207.

Schlötter, P., 2005. *Vertraute Sprache und ihre Entdeckung: Systemaufstellungen sind kein Zufallsprodukt – der empirische Nachweis.* Carl-Auer-Verlag: Heidelberg.

Soon, C.S., Brass, M., Heinze, H.-J. and Haynes, J.-D., 2008. 'Unconscious determinants of free decisions in the human brain'. *Nature Neurosci.* 11, 543–545.

Varga von Kibéd, M. and Sparrer, I., 2014. *Ganz im Gegenteil. Tetralemmaarbeit und andere Grundformen Systemischer Strukturaufstellungen – für Querdenker und solche, die es werden wollen.* Carl-Auer-Verlag: Heidelberg.

Wahr, F. and de la Harpe, B., 2016. 'Changing from within: an action research perspective for bringing about sustainability curriculum change in higher education'. In: M. Barth, G. Michelsen, M. Rieckmann, and I. Thomas (eds.), *Routledge Handbook of Higher Education for Sustainable Development.* Routledge: Oxford, 161–180.

5 Systemic constellations in organisational and institutional contexts

Systems constellations provide good applications in systemic and evolutionary contexts. Strategy and innovation processes, as well as concepts in the light of a sustainable development, can be analysed on the basis of evolutionary-systemic approaches. Evolutionary-systemic approaches provide a different view of organisations and allow ways to better understand the learning and development processes of social systems. Transformation, change and complexity are structural elements examining strategies and innovations for their contribution to a sustainable development. Both evolutionary and systemic conceptions relate to change and complexity, as well as to the object of investigation with each other, and also highlight processes as central issues (Osterhold, 2002). A basic assumption of both approaches is a dynamic, complex and constantly changing world. System-oriented concepts emphasise the self-organisation and self-dynamics of systems. The special gain of systemic approaches is based on a different point of view because they can demonstrate phenomena presented in a shortened, insufficient way or not at all in a linear-causal view. Systems theory focuses superficially on self-organisation, patterns and complex structures, in addition to emphasising the relationship of the single elements among one another (Bestehorn, 2001). In that sense, the quality and the specific nature of the relationship are of particular interest. The latter can be described as interactions and shows that system elements do not exist in isolation but influence one another, and thus a so-called recursivity arises (Vogd, 2005). Interaction processes are therefore circular, or they or reverse and complicate or hamper clear cause–effect classifications. The clear significance of systemic perspectives can be seen in the superposition of several logical chains and levels of description.

In general, it is crucial to distinguish different levels having an influence on the current situation or problem as well as on the solution. As in business contexts, people are part of the main transaction and exchange processes; there is always a personal level that plays a role. Yet beside the personal issues, there is an organisational level that constellators should always bear in mind, as well as the more complex system level (compared to family systems), because there are much more moderating factors and interrelations. Guiding questions for each level can be:

- *People level:*

    - How is the situation influenced by the personal patterns of the employees having their origin in the history of their life experiences or dynamics in their families?
    - Are these patterns possibly re-enacted within the organisation, or how do these patterns influence organisational events?
    - Do relationship conflicts or problematic patterns of communication among employees or between departments have an influence?
    - . . .

- *Organisational level:*

    - Does the organisation have functional structures? Could recurring relationship difficulties express fundamental problems of the system?
    - Do all people have their appropriate place?
    - Are leadership and management responsibilities practiced adequately?
    - What is the role of management systems?
    - How is innovation managed and integrated within the organisation?
    - How about culture?
    - . . .

- *Systems level:*

    - Have there been any changes in the environment (e.g. market, competitors, customers and stakeholders) that the organisation has to adapt to or deal with? Are any changes expected?
    - What about the history or the founders of the company?
    - How are cultural differences managed and integrated (within multinational companies)?

- Are the existing management systems functional and appropriate for the whole value chain?
- How is change managed? How is transformation dealt with?
- ...

These questions can be helpful for clarifying the concern as well as for testing further interventions. In addition, during a constellation, several contexts and structural levels can become important and resonate ongoing. In tetralemma constellations, diverse levels often become obviously as decision making is also influenced by different topics. For instance, stressing sustainability education in a tetralemma constellation, *this one* could relate to a holistic or presidential perspective, *the other* could be more related to faculty issues, *both* with a whole-institution approach, and *neither* could focus on learning techniques. There is a permanent resonance of different levels within constellations, so the constellator can induce for certain interventions or work with spontaneous changes of structural levels.

Main systemic principles

Beside different structural levels, there are several guiding orientations and principles in systems work. Sparrer and Varga von Kibéd (Daimler, 2014) distinguish a general systems orientation, principles of systems orientation and systemic principles of compensation. They can help to find solutions and to test interventions; thus it is likely to find them in systems, but they do not have to necessarily capture a system. According to the authors, there are four main general systems orientations (Daimler, 2014: 40 ff.):

- *Securing existence:* This is knowledge of boundaries and who belongs to a particular system, like family, organisations, body systems and so on. So organisational belonging terminates when leaving the organisation.
  This makes a real difference in working with systems because different underlying dynamics have an influence on improvements, progress and satisfying solutions.
- *Growth and reproduction:* Establishing good conditions is crucial, in that growth is important for a system. So seniority within the system is important as well as the appreciation of the older parts of any item, element and so on. Moreover, it is essential to recognise that some systems cannot grow further, like founders or incorporators or a group of survivors.

When reproduction becomes important in creating systems of comparable manner, like family or subsidiaries, the boundaries, beneficial terms, and conditions should be clarified and established. In early relations but limited for a certain period of time, the older system should invest more resources into the new, upcoming system vis-à-vis comparable systems. Daimler (2014: 41) describes the following examples: multinationals creating a new subsidiary or parents having a new baby should provide more resources for a certain period of time to the new system compared to the existing ones, like other, older subsidiaries or siblings. However, boundaries as well as space for one's own development are crucial for the upcoming new entity.

Discussions about distribution, allocation and justice should have this orientation as a guiding principle in mind. In academia, this principle is sometimes out of sight when implementing new concepts, faculties and so on. However, it is important to define periods of time and boundaries for this particular and extraordinary support.

- *Boosting the immune system:* Long-term-oriented systems develop principles and functions comparable to the immune system in order to survive and react to irritations and attacks appropriately. According to Sparrer (2016: 51), in social systems, the ability to communicate, bearing responsibility and commitment are crucial elements of immunity. She argues that the better the members of a system can communicate with one another, the faster new information can be processed and spread. The greater the commitment of the members is, the better they can overcome times of crisis.

As in the whole-institution approach, academia has a central role in society as well, which has to reflect the communication ability of academia. Fostering sustainability as well as progress in society, academia needs new ways and forms of communication and dissemination of new knowledge to serve other systems and boost the immunity of society as a whole.

- *Individuation:* In the event that a system wants to develop, learn and change, the appreciation and facilitation of individual capabilities become pivotal. So individual capabilities, specialisation, promotion as well as creative processes should be enabled. Systems not able to change and develop themselves further are often limited to the constraints of individual capabilities.

With the system orientation of individuation in mind, there will be different ways of facilitating sustainability in or by academia. However, in fostering sustainability in academia, there are different individual capabilities of academic members and thus of the organisational capability to facilitate sustainable development and change. So a conceptual framework for academic sustainability or sustainable academia should always provide several degrees of freedom for individuation.

Sparrer (2016) and Daimler (2014) also stress that the guiding systemic principles can be helpful in prioritising processes and interventions but are no indicators of importance. In detail, the authors amplify specific systemic principles, like

1   the principle of not denying or recognising what is;
2   the principal of belonging or affiliation;
3   the acceptance of chronological order;
4   recognising greater commitment for the whole; and
5   the priority of higher performance and capabilities.

The first principle is important for any change and real transition. It is about the recognition of what is there, what is given. It addresses an attitude encompassing processes and activities for initiating change and implementing actions. The overview of what is given and the recognition of the status quo are the basis for all ongoing decisions. This basic analysis also comprises the people level (organisational members, team values, students, etc.), the organisational level (corporate value and identity, projects, etc.), as well as the systems level (customers, suppliers, stakeholders, etc.), alongside business indicators.

The second principle enables security and stability inasmuch as belonging and affiliation enhance transparency and get things straight. In organisations/institutions, a lack of clarity often leads to the absence of liability and thus limited productivity (Daimler, 2014). Useful questions can be aimed at founders, old or former members or projects, the consideration of old and new values and so on.

The third principle is a curative but not normative one for growth-oriented systems because the issue of who belongs longer to the system and who is richer in organisational experience is of importance. In growth-oriented systems, longevity and experience take priority.

The fourth principle stresses the importance of recognising greater commitment for a system because it boosts immunity. Both overemphasis and negligence devitalise a system's immunity for future crises. The lack of clarity is overwhelmingly energy-sapping. Hidden hierarchies can emerge and cause infighting, intrigues, inner notice and other issues.

Final, the fifth principle highlights the ability to change, develop further and remain competitive. Therefore, a focus on the capabilities of the organisational members is necessary. An appropriate organisational culture, fostering creative processes as well as accepting mistakes within the change processes, can support the development of capabilities.

Principles of compensation, balance or adjustment

Moreover, systemic work is also strongly related to a degree of give and take. The term balancing or compensation is not used objectively in systems work. The necessity for compensation often depends on the need of the respective person. Sparrer (2016) stresses that compensation is strongly related to the construction of one's own reality, world view and perspective. It is very comparable to the phrase about whether the glass is half full or half empty. Thus, constellation work enables the reframing of hidden constructions and comes up with new solutions. The principles of compensation enable people to notice as well as to recognise emotional, immaterial or material imbalances. Interestingly enough, in transferring ethical conceptualisations of guilt or innocence in economic contexts, solutions can be found more easily (Sparrer, 2016: 46; Daimler, 2014: 54). Here are the principles of compensation:

1 Increase balancing in good.
2 Decrease balancing in bad.
3 Avoid too precise balancing.
4 The debtor has a right of instruction or reminder.
5 The creditor violates the right the debtor if he refuses to remind or instruct.
6 The debtor's balance should be affected by the creditor's 'valuta'.
7 The real balancing is in the recognition of the compensation obligation.
8 Compensation will be effective only by denoting this recognition.
9 Rejecting compensation annuls the already given recognition of the compensation obligation.
10 Mixing the balance with the compensation cause the compensation to degenerate to a plain payment.

The following are brief explanations of these principles:

1 It is useful and meaningful to emphasise convenient actions and behaviour to keep commitment, giving and generosity going. Recognition can happen by words, celebration, award, premium or the like.

2 This is about fault tolerance, the avoidance of equal revenge or an eye for an eye, a tooth for a tooth culture. Tit for tat should always be meant to restore cooperation. It does not mean that consequences are possible but in a meaningful manner and always emphasising the ability to learn; otherwise, cover-up will increase and only weaken the organisational immune system.

3 This principle is only for relationship-oriented systems or in cases where relations are important and relationships should be sustained. The exact calculation and settlements in relationships cause more or less separation, whereas generosity in thoughts and actions leads to more enrichment and other qualities of relationships. Daimler (2014: 56) also states that there are differences in the measures and arrangements of companies or entrepreneurs, depending on the level of customer relationship and loyalty. Shops in holiday areas or in roadway stations or even ice cream salespersons on beaches have less relationship-oriented tools compared to other stores. In academia, there are also differences in the orientation of relationships. Relations between professors and students are not often long-term oriented. Even scientific staff are often employed only fixed-term. Recent discussions about participation in academia address the limited period of time students belong to the system. Thus, here is a great difference between academic people, administration, research, teaching and stakeholders.

4 This is a helpful systemic principle as it enhances the likelihood of getting what someone wishes. When taking over something from a colleague or institutional section, this person should talk about it and ask for compensation. In addition, it is important to remind someone to compensate in case the person forgets.

5 This falls in line with principle 4 and goes beyond it because it emphasises the act of reminding as a discrete duty. If it is missed, the compensation will be reduced.

6 There are often differences in the idea of how or what to compensate, and the types of compensation are mostly only partly convertible. Simon (2007) compares this notion with differences in bookkeeping concerning the giving and taking of different people. Compensation is thus effective only when transferred into the addressee's language and valuta. Therefore, appropriate compensation should be always asked for.

7–8 Both principles are well connected. It is a mistaken idea that compensation service (e.g. money transfer, food, etc.) itself is compensation. Compensation is effective and solution-oriented only if the debtor (or culprit) acknowledges and recognises his own compensation obligation. Otherwise, different levels will be mixed and hinder good solutions.

9   In order not to depreciate a compensation, it is necessary to just recognise the compensation obligation, even without effective compensation being paid.

10   This principle addresses cases in which one person asks the other how to compensate and then just transfers the compensation. However, just transferring the desired compensation escalates relations further because the transfer is not combined with a real recognition but with heedlessness.

Differences between family systems and organisations/institutions

It is important to state that there are differences between family and organisational systems. The key dynamics of family systems cannot be easily transferred to organisational or business systems. There are differences in underlying dynamics, as well as in mediating or moderating elements. According to König (2007), the following aspects have to be considered in particular when working in a business context, as well as in research and teaching:

1   *Belonging or affiliation and exclusion criteria:* Organisational affiliation is acquired temporarily and may be terminated by either side. People are basically exchangeable, and just that causes the transpersonal continuity of organisations. In organisations, the focus of people is set more on roles and functions, whereas it is more holistic in families. Yet family and work influence mutually, and new ways of living and working even invert common logics.

2   *Seniority within the system and compared to other systems:* In family systems, there are vertical and horizontal structures, hierarchical and temporal logics. Horizontal orders cannot be changed in family systems, whereas vertical structures can be changed by starting a new family. In companies, vertical structures and positioning regarding seniority are temporary and focused just on the professional role of a person; positions are assigned, not given naturally. There is a greater focus on capabilities and experience in organisations or institutions. Moreover, lifelong affiliation is not appropriate for every group, and there is also change in working conditions, so different time periods of affiliation become more crucial. König (2007) stresses a rigid

handling of seniority as it is more related to resistance to change, and this leads to resource allocation conflicts.

3 *Hierarchy of positions and roles in an organisation:* Hierarchies are influenced by different aspects in organisations. Leadership, responsibilities, duties, competency, experience and time of membership within an organisation are of high importance in management contexts. Ambiguities always cost energy. Hidden hierarchies should be avoided. In addition, there are further micro-, meso- and macro-level effects influencing the organisational reality. If leadership is not perceived or not adequately perceived, a variety of problems and problematic situations will be caused. Currently, more complex and flexible forms of leadership are necessary in order to meet present challenges. Thus, hierarchical concepts do not often meet these requirements. In addition, there are changing balances of power, so a specific and open view is necessary in constellations.

4 *Modalities of giving and taking and related power relations, relationships and obligations:* Altruism and love are the main system logics of families. The central logics in organisational contexts are profits and wages (money). The balance of giving and taking is more related to family, whereas the main compensation in an organisation is based on bartering or countertrade. This kind of exchange or justice can be done over several generations in family systems – an idea anchored within the externalities of economic activities as well.

5 *Commitment and performance and their recognition – the more engagement, the more one has to say:* Organisations have to face change permanently; thus they should honour special engagement by their members. Therefore it is of high importance to assign performance where it is generated. Otherwise, the long-time performance of members has to be honoured as well. Yet the priority of higher performance and capabilities has to be considered.

6 *Functionality internally and externally, the latter awarded a more dominant position:* In business contexts, it would make more sense not to give certain notions of orders in constellations but to use constellation work to raise and to improve the understanding and functionality of leadership and the like on the basis of the organisational members and the given information. König (2007) highlights that the difference between inside and outside is just one relevant distinction among others. He also strives to develop further distinctions to make the situation and consequences visible and noticeable (in order to find solutions, enhancing the space of options).

7   *Transience. If an organisational system loses its task, there must also be the possibility of resolving it:* There is a big categorical difference concerning time in the family and the organisation. The temporal order of the family is therefore directly coupled with our lifetime, whereas organisations do not have this limitation. The two types of systems have different time horizons. König (2007) emphasises the decoupling of the two different systems within constellation work.

In summary, several guiding orientations and principals are useful in order to find solutions and develop other different types of systems. They are heuristic, and they appear regularly in constellation work in different ways. High-quality constellators have these principles in mind and are open for everything that arises and emerges within a systemic constellation (see also eight valuable assistants, Figure 2.1, and Chapter 2). Moreover, constellators are not omniscient but make suggestions based on the principles serving the process of finding solutions. Academia seems to be a very special institution as different types of people work within this institution having different types of limited or unlimited contracts. Trade-offs seem to be inherent. Consequently, systemic structural constellations provide a good method to make different system levels clear, to separate mixed issues and to clarify relationships in order to find appropriate solutions for academic challenges.

References

Bestehorn, M., 2001. 'Musterbildung und Musteranalyse – Moderne Aspekte der Synergetik'. In: A. Gnauck (eds.), *Systemtheorie und Modellierung von Öko-systemen.* Springer: Heidelberg, 18–32.
Daimler, R., 2014. *Basics der Systemischen Strukturaufstellungen.* Kösel: München.
König, O., 2007. 'Aufstellungsarbeit zwischen Supervision, Beratung, Therapie und Ideologie'. In: *Gruppendynamik und die Professionalisierung psychosozi-aler Berufe.* Carl-Auer-Verlag: Heidelberg, 150–176.
Osterhold, G., 2002. *Veränderungsmanagement: Wege zum langfristigen Unternehmenserfolg.* Springer Gabler: Wiesbaden.
Simon, F.B., 2007. *Einführung in die systemische Organisationstheorie.* Auer: Heidelberg.
Sparrer, I., 2016. *Systemische Strukturaufstellungen: Theorie und Praxis.* 3. Aufl. Carl-Auer-Systeme Verlag GmbH: Heidelberg.
Vogd, W., 2005. *Systemtheorie und rekonstruktive Sozialforschung: eine empirische Versöhnung unterschiedlicher theoretischer Perspektiven.* Springer-Verlag GmbH: Opladen.

6 Systemic constellations in academia in practice

Academia is a very unique and special system, for which constellation work can focus on several aspects and has to consider manifold principles as well. As described in Chapter 5, there are several time frames depending on the group of people that is focused on, such as professors, deans, presidents, students, administrators, stakeholders and others. Moreover, constellations can focus on academia itself and its sustainability impacts or on its contributions to a sustainable development. In this chapter, different constellations are shown, highlighting the huge possibilities for application in academia. The first example is from research, the second one from teaching, and the third from institutions. The last two examples present possibilities for experimentation and opportunities for supervision.

Research and consulting/transfer

This case is classified in the research and consulting, or transfer, area as it was a joint work with a company representative. On the one hand, there was a particular concern; on the other hand, the interested group of students and researchers wanted to further investigate the industry to which the company's representative belonged. Therefore, half a day was organised to conduct several systemic constellations. First, the particular concern of the company representative took place. It was a real-world problem, or challenge, illustrating a specific example from the water industry. By means of a systemic structural constellation, it should be clear how network engagement aimed at sustainability could be fostered and how the companies or the network should be addressed or focused on. The constellation process followed the rules described in Chapter 2 and was organised like this: Firstly, the concern was clarified and relevant elements were selected on a joint basis. Then the issue-holder selected representatives and placed them in the room. All representatives had the possibility to speak about their emotions and impulses to move. Water was set as a free element moving on his own impulses. All other representatives were asked to move or to signal their wish to move or change position. After a structured start, the elements were allowed to talk to one another to strengthen the dynamics. Within the process, the fear of change indicated that the companies have to initiate firm internal change processes as well. The representatives signalled that there has to be a clear positive change for the respective companies. Sustainability was introduced as a further element inasmuch as there were mixed contexts concerning the network idea, water and sustainability issues. Integrating sustainability within the constellation was a great relief for most representatives. Every discussion and movement showed that a shared mission was missing, so that it would not be a good idea to found a new network promoting a good idea with selected partners in the beginning. During the process, it became clear that all companies had a different idea for the network initiative and its goals. So, again, the best way to start with the network was not the establishment of the network with selected partners based on a given idea but rather to start a communication process concerning the role of water for the companies and the embedment of this pure resource in the context of a sustainable development (see the initial and final constellations stressing highlights in Figure 6.1).

The final constellation shows that there is still a different orientation of the big water companies concerning change, water management and sustainability, so they have to clarify specific aspects with one another. Sustainability is more focused on the small water companies and the network idea, meaning that the original idea and small companies should clarify their understanding of sustainability and develop a clear vision. The fear of change is still among big water companies, so they have to change on an organisational level as well. Final, the whole process was reflected and discussed with the attendees. In this example, the application of the systemic structural constellation and the cooperative reflection process prevented the initiators from the spread of a Trojan horse and enabled them to gain new insights of network formation and cooperative engagement. Moreover, it became clear that mixed contexts have to be separated to enable effective management and networking.

This constellation is a good example of using the power of unconscious knowledge and systemic relations. The gained insights can also be reflected on to see how to foster sustainability within academia. Starting initiatives to foster sustainability within academia or on the basis

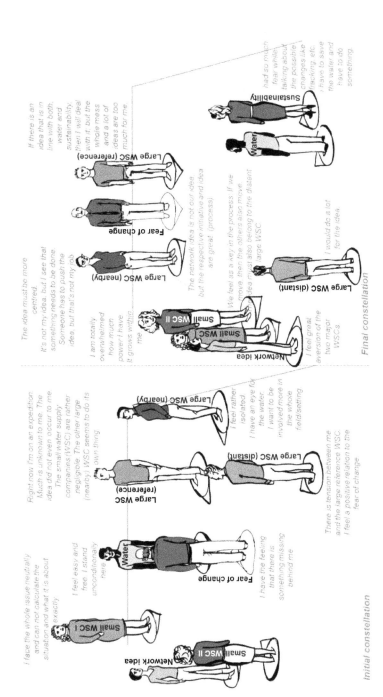

Figure 6.1 Example I of constellation in research and transfer

of a whole-institution approach requires a clear vision, but it is likely that there are different understandings and orientations among particular groups in academia. Thus, sustainability goals should be discussed within academia as a whole and among all different institutional members in order to find a joint understanding and appropriate starting points for activities. In light of a whole-institution approach, all members should be integrated within such dialogues, and the different times of the affiliations of specific groups should be reflected and discussed within such processes as well. Even stakeholder integration could be a good idea inasmuch as how they expect to make academia more sustainable is of importance.

As the specific content of the network idea was not fully clear, even after discussion, three different content orientations of the idea were tested: (1) pure water, (2) technical progress and (3) social development (see Figure 6.2, left side). This constellation was organised in a partly inconspicuous way, so the three content orientations did not know what they represented in particular. The network idea was free to stand in front of the different three options. The first impulse of the idea was to go closely to the option 1 (pure water), although there were good relations with the other two alternatives. In this sense, it was a good addition to the first constellation and a clear and appropriate starting point for developing the network idea into a network on a joint basis of different partners and visions.

In addition, setting a deeper focus on research, further insights into the resilience and the ethos of the water industry were figured out, together with the company representative. Further working on a systems level, a matrix was used formed of transformation and disorganisation as one centre and of growth and embedding as the second centre, all four centres represented by persons. Several elements, in particular the water company, clients, ethos, national competitive company, international operating water company and so on (see Figure 6.2, right side), were set hidden and asked to find their way within or around the matrix. The elements were placed one after another within this matrix. Then the constellator asked for their representative perceptions, impulses and differences. This constellation work showed that the water industry is well established and focused on growth and transformation. In addition, companies have a strict cost orientation, want to maintain structures and infrastructures and do avoid wanting to overload their companies with diluted sustainability issues. Yet the deep structure of the industry is not really transparent or conscious. The ethos of the water industry is closely connected with pure and natural water as one of our most crucial resources. Several assumptions and hypotheses were formulated for further research.

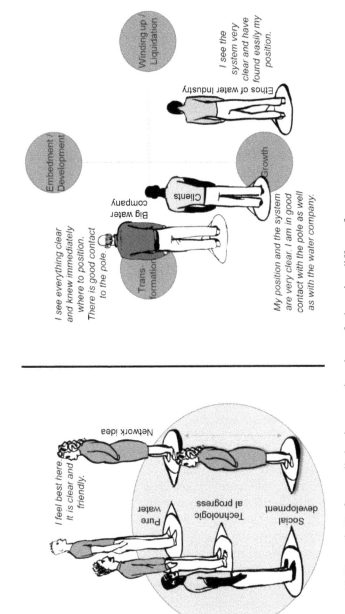

Figure 6.2 Example II of constellation in research and transfer by using different forms

This example of research is not only bonded to particular research interests. The insights can also be transferred to academia asking and reflecting how academia can secure good-quality water, avoid contamination and look for alternative resource flows in the context of water usage and wastewater, recycling opportunities or closed water systems. So constellations in a whole-institution approach can send impulses from one academic area to another and foster both sustainable academia and academia for a sustainable development. In this case, several ideas for research, consulting and teaching emerged and were integrated in ongoing constellation work. In this sense UNESCO (2014: 171), this is a new form of research, providing high levels of participation as well as multi-stakeholder social learning.

Teaching

Another example is from teaching (see Figure 6.3). Constellation work can easily be integrated in traditional lecture settings, as well as being offered as block seminars. In case of teaching, the students get an idea of other methods focusing more on unconscious information or knowledge and visualising systemic relations. In light of whole-institution approaches, students are directly integrated within the learning process and can even take the part of initiators for research-based questions and project work. Systemic structural constellations emphasise interactive, integrative and critical forms of learning (UNESCO, 2014: 171). The given example is about a lecture unit in the topic of risk management working together with international academic managers. All were asked to develop and implement a particular project within their home university or academic institution. Using system constellation, the main goal was to find out whether there are implicit or hidden risks that the academic managers do not see or estimate.

Therefore, six people were selected on a voluntary basis and asked to represent diverse risk forms, like strategic, environmental or governance risks. The lecturer and constellator allocated the respective risks and asked the representatives to take positions within the given space. In a second step, some managers represented their ideas and stepped in front of the respective risks. They had some time to share their feelings and differences and to give feedback to each other before moving on to the next risk. There were some surprising effects, like circling around the governance position because there is no safety and security for the project, academia or even people (due to the instability of regional development).

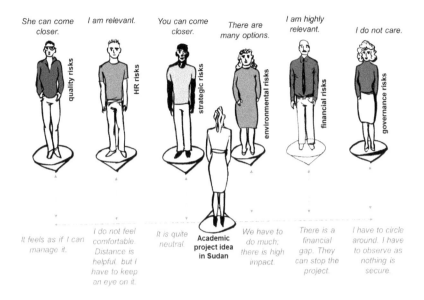

Figure 6.3 Example of constellation in teaching – testing options for projects

Moreover, the academic managers were impressed by getting additional information for implementing their project successfully within their institutions, as well as different levels having an impact on projects. Afterwards, new insights or additional information can be reflected on and integrated within the respective project in order to make it more effective.

Institution and administration

A third example is based in the institutional area (see Figure 6.4). The group wants to foster sustainability within academic life but is not sure about the best way to begin their initiative. So issues-holders were the whole group of people selecting a head for the constellation. After a short initial conversation, six different representatives were selected and placed in the room. The initial constellation faced a difficult situation to start with because it was challenging to address the idea of more sustainability within academic life. Sustainability did not really feel integrated and was situated outside the system; however, within the process, it became more and more central. As it was unclear how to position sustainability within this system, several positions were tested, and finally the vice president II

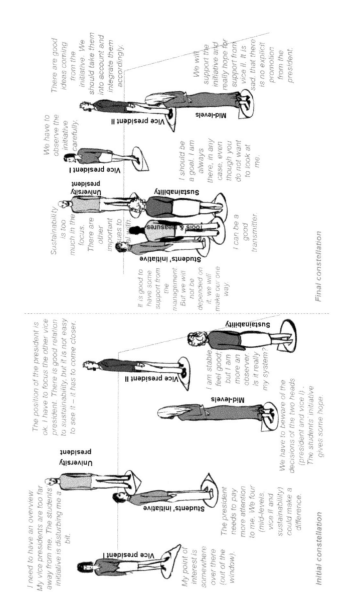

Figure 6.4 Example of constellation in an institutional context

Initial constellation

Vice president I

My point of interest is somewhere over there (out of the window).

University president

I need to have an overview. My vice presidents are too far away from me. The students initiative is disturbing me a bit.

Students' initiative

The president needs to pay more attention to me. We four (mid-levels, vice II and sustainability) could make a difference.

Mid-levels

We have to beware of the decisions of the two heads (president and vice I). The students initiative gives some hope.

Vice president II

The position of the president is ok. I have to focus the other vice president. There is good relation to sustainability, but it is not easy to see it – it has to come closer.

Sustainability

I am stable feel good, but I am more an observer. Is it really my system?

Final constellation

Vice president I

We have to observe the initiative carefully.

Vice president II

There are good ideas coming from the initiative. We should take them into account and integrate them accordingly.

Mid-levels

We will support the initiative and really hope for support from vice II. It is sad, that there is no explicit promotion from the president.

University president

Sustainability is too much in the focus. There are other important things to deal with.

Sustainability

I should be a goal. I am always there, in any case, even though you do not want to look at me.

Tools & measures

I can be a good transmitter.

Students' initiative

It is good to have some support from the management. But we will not be depended on it. we will make our one way.

had the idea that there is a kind of transmitter missing between the initiative and sustainability.

The representatives discussed the necessary tasks of this transmitter, but the respective representative had a clear idea of what is needed and what he represented: He argued his case for inventory or a survey concerning the steady-state situation of academia. The representative said that tools and measures are needed for imposing data and representing responsibility. Talking about corporate social responsibility (CSR) – within the initiative group, tools and measures, sustainability, mid-levels and vice II – showed that CSR is just a transition status or passage towards sustainability. The final constellation made obvious that even just a bottom-up initiative activates some changes towards more sustainability within the academic system. Reflecting and discussing the results of the constellation, it became clear to the group that there are different levels, time frames, tools and roles for supporting sustainability in academia. In this case, it was also discussed whether it is worth initiating a university-wide roundtable stressing sustainability issues and integrating interested stakeholders within the topic or how to establish alliances to foster sustainability. The group came up with diverse ideas; some of them were developed further by means of systemic constellations.

Experimenting and supervision

Figure 6.5 emphasises the work with artefacts as one crucial process facilitator in constellations, emphazised by Jürgen Rippel working at Ansbach University of Applied Sciences. From a content perspective, the constellation focused on strategic intuition (as described by Duggan, 2013). Several elements were selected and assigned in a hidden way. They got time to find their place within the matrix and were then asked for their impulses and differences. In a second step, they got the opportunity to choose an artefact. After choosing an artefact, the expressions become much clearer and pointed. Either the relations were clarified, or the location within the matrix became more obvious. In case of the element foreshadow, something was blocking the element in the beginning. After wearing the mask, the representative lost this blocking feeling – as was represented by the mask – and could speak more clearly and more precisely. Moreover, the element strategic intuition could formulate its location better and became aware of its own function and characteristics within the matrix.

Thus, constellations can be used to experiment and to gain new insights or simulate specific interventions in a visual and spatial 3D depiction,

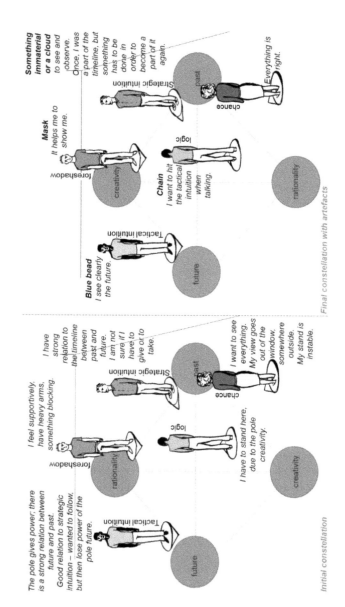

Figure 6.5 Example of work with archetypes

stressing relations, immediate changes and challenges. Experimenting with sustainability is one core issue in academia among others, like quality, diversity, internationalisation, transfer and the like. It became clear that sustainability is one issue among other issues that the Steering Committee has to deal with and that this is an emerging new topic, so it will take up room. Referring to the previously listed systemic principles, it is necessary to recognise that other issues are older and that this new topic is competing for resources. In this regard, sustainability has to become a top-down topic. Otherwise, it will be not accepted within the whole institution.

Given the systemic constraints of implementing sustainability within academia, supervision constellations are an excellent format for analysing those constraints and for simulating interventions and transformation stimuli, in preparation for the next steps. Within these constellations, it can become obvious, from a systemic perspective, what has to be recognised, integrated, added or disengaged in a given context.

All in all, in complex contexts such as sustainability and resilience, learning progress can be achieved in all academic areas and on the knowledge level as well as on the level of action. So transdisciplinary research processes can be rapid problem identification and problem structuring in order to allow direct problem solving in cooperation between science and practice and to enable direct transdisciplinary integration (Hirsch et al., 2008). In teaching, the method enables high levels of participation and innovative interactive, integrative and reflecting forms of learning. From an institutional perspective, a multi-stakeholder approach will be likely and will foster cross-level discussions.

References

Duggan, W., 2013. *Strategic Intuition: The Creative Spark in Human Achievement*. Columbia Business School Publishing. Columbia University Press: New York.

Hirsch, G.H., Hoffmann-Riem, H., Biber-Klemm, S., Grossenbacher-Mansuy, W., Joye, D., Pohl, C., Wiesmann, U., and Zemp, E. (eds.) 2008. *Handbook of Transdisciplinary Research*. Springer: Heidelberg.

UNESCO, 2014. *Shaping the Future We Want*. UN Decade of Education for Sustainable Development (2005–2014) Final Report, Paris, France.

7 Applications for and in academia

Systemic constellations can be used by people, organisations, science and other areas to focus on diverse issues highlighting solutions, reflexions or options, like communication, team motivation, restructuring, decision making, strategic issues, ethical questions, and sustainability conflicts or dilemmas. Representations can be used for revealing new perspectives on all the issues addressed, like business, politics, religion, cultures, war and crimes, philosophy and the like. Systemic constellations are an effective way to teach complex relationships and multilevel challenges and to learn how unconscious or tacit knowledge can be made tangible and visible. There are several advantages to their use in teaching, research and institutions. The method appears to be a very powerful tool in explaining and transferring multicausalities in systems and can be complemented by offering itself to traditional methods. Findings, implementation options, conclusions and other results from work with systemic constellations are often not attainable by a pure study of documents, interviews or an empirical survey, at least not at a comparable speed. Today's topics and studies have to be taught and educated more and more in multicausal and inter- and transdisciplinary contexts that require new and innovative methods. The integration of systemic structural constellations in research and teaching allows the teaching and learning of complex relationships, multilevel challenges and sustainability transdisciplinarily and makes the importance of sustainability for businesses tangible and visible. Thus, it provides essential knowledge for social contexts. Table 7.1 highlights several advantages in the main academic areas.

In teaching contexts, a main progress is the participatory learning context inasmuch as students will increase competencies by using more senses than previously and by activating their own unconsciousness. They will learn to see, perceive and understand systemic and circular relations,

Table 7.1 Effect of systemic structural constellations in the fields of teaching and research and in institutions

Use in teaching	Use in research	Use in institutions
Arrangement of systemic competence	Search for structures and patterns	Structuring of challenges
Visualisation of contents	Relationship analysis	Focusing on questions and solutions
Spatial language as a quick orientation	Illustration of stress fields	Visualisation of systems components and structure
Action learning	Knowledge of the deep structures of systems	Expatiating hidden or unconscious knowledge
Connecting science and practice	Changing patterns of knowledge	Simulation of interventions and different options
Motivating learning contents by diverse senses	Forming hypotheses or research propositions	Reflexion of functional and dysfunctional elements
Supervision	Testing and simulation of interventions	Consulting, coaching and supervision

relationships or complex paradoxes better. In teaching and raising the awareness of complex issues, such as sustainability and resilience, a learning progress can be achieved on both the knowledge level and the level of action (Parker et al., 2004). Transdisciplinary systemic thinking can be developed quickly into a system by the representation of an element. In learning settings, systemic structural constellations can be an additional tool in sustainability pedagogics beside stimulus activities, future visioning, case studies, critical reading, group discussions and the like (Evans, 2016). It is a setting for joint – both lecturers and students – learning, investigation, exploring, amazement and advancing concepts or tools.

In the research context, multi-stakeholder integration is enabled. Researchers can analyse phenomena from another perspective and integrate hidden, tacit or unconscious knowledge within their investigations. The possibility to involve community is crucial. Moreover, systemic structural constellations enable transdisciplinary collaboration and thus the integration of external stakeholders within learning contexts (Barth, 2015). So working with constellations, different types of research are possible, like basic research, used-inspired basic research, as well as applied research. Either sustainability topics can either be improved by means of constellation work, new assumptions or hypotheses can be formulated for further research or specific relations or ideas can be simulated to foster or renew sustainability theories and concepts. It is also possible and useful to integrate all options in research-based learning contexts.

Institutional contexts enable the revelation of hidden barriers or the necessary recognition of underlying dynamics. In all cases, momentum can take place and be stimulated by diverse interventions. Involving external stakeholders as issue-holders can also strengthen leadership and motivation by jointly working on sustainability challenges. Staff integration or participation can be easily managed because they can be asked to take part in a constellation session either as representatives or as observers and discussants. A step-by-step engagement can be easily developed and fostered.

As already mentioned, the findings, insights, implementation options, conclusions and other results from work with systemic constellations are often not attainable by a pure study of documents, interviews or an empirical survey, at least not at a comparable speed (a constellation takes 0.5–2 hours). Additionally, the experience of a system image and the corresponding effects and relationships can make a difference in perception, evaluation and future action. Thus, transdisciplinary or multi-stakeholder research processes enable rapid problem identification and problem structuring, allow direct problem solving in cooperation between science and practice and permit the direct transdisciplinary integration in research contexts (Hirsch et al., 2008). However, when introducing new methods in working contexts, resistance and constraints always have to be overcome. Most students and people have never experienced systemic constellations before, some have had a bad experience or others do not want to experience the process but just to observe it. So the constellator or facilitator has to accompany persons patiently while they get to know this new method and become part of it. According to my experience, four aspects have a major influence on how easy the beginning and the willingness to take part are: (1) group or group dynamic, (2) culture or geographic region, (3) adventurousness or innovativeness and (4) age or maturity. If people do not believe that the method is functional, it is helpful to work with hidden elements so that people can observe and realise that it is not comparable to role-play or theatre, but something different.

Some experiences

- Representing dead persons or dictators in a hidden constellation can have a strong effect as the persons experience remarkable differences between their own feelings and the representative perception. Representing dead persons in a constellation reveals completely different feelings compared to representing those who are currently alive. A student was aghast after realising that she represented a dictator (in our Western interpretation and evaluation) and said, "Oh no, I would stand there or I would react and act differently". In this case, two issues are to be highlighted: (1) It is useful to start with hidden representations as they are more convincing of the method's performance, and (2) representative perception works best when people do not govern constellation processes by using their mental, logical power. The latter can be trained.

- Starting with daily experiences like preferences for food or outfit, sports or books is likely to convince people of the method's performance in a first contact. So different types of nutrition can be distinguished, like meat, vegetable, vegan and so on, or types of outfits – sporty, chic, functional among others. Here it is also helpful to start with hidden constellations.
- If students do not want to take part in a constellation, it is helpful to start with the interested ones and ask them to take part as observers, a commentators or researchers, so that they can join as well and contribute to the constellations by their observations and analysis of differences to already known relations or approaches.

Asking students for their feedback regarding the integration of systemic constellations into teaching, the following answers are quite representative.

- "It is great that we have the chance to try alternative methods."
- "The method is a great supplement to the traditional methods of economics."
- "My other focus is controlling, so I was very sceptical; yet there is a lot of fun and it is incredibly informative."
- "I cannot understand why the method is not more widely spread."

One student even got a job offer in marketing/market research precisely because he used the method of system constellation in his study programme. Company representatives are all positively surprised by the possibilities and the substance of insight gained by the method. They experience the joint reflection as enriching.

However, the enthusiasm of the students and practice partners for the method of systemic constellations cannot hide the responsible use of the method. Its use in transdisciplinary research and teaching calls for sound training in systemic structural constellations and a respectful interaction with people and their concerns. Just because the method has the potential to enable innovative teaching, research and practice for identifying and initiating new transdisciplinary options and ways for action and implementation, an ethical and respectful treatment of concerns, people and processes is mandatory. Moreover, the lecturers' role changes as new content-based issues are discovered jointly. So the knowledge gap between lecturers and students becomes smaller, and lecturers lose their unique position of being experts. By using the method in transdisciplinary contexts, the responsibility can go far beyond the university system, which might be the reason for the large-scale scepticism and awe of systemic constellations. Anyway, past experience shows that the unity of research and teaching in transdisciplinary contexts should be pursued.

However, system constellations do not necessarily cause systemic and transdisciplinary action. Beside the revelation of hidden or non-visible patterns, the mesh of relations and interaction structures and the subsequent discussion with company representatives and students are of particular importance for the transdisciplinary teaching and research process. The visualisation of depth structures and hidden patterns allows for a discussion of science and practice at eye level, as well as a for the mutual stimulation of theory, practice and reality. Figure 7.1 provides an example for integrating constellations within teaching–researching contexts. Starting a lecture, both an introduction to the method and, in particular, to sustainability topics (e.g. water, management tools, green innovations, etc.) should always be given. Before a specific sustainability constellation is started, students should get to know representative perception. Here, the previous examples can help. Then some further constellations can be conducted before integrating a specific concern of stakeholders or company representatives. In accordance with whole-institution approaches, it is highly recommended to work jointly with students, researchers and stakeholders – as fruitful insights are possible for every group. After a constellation, a mutual discussion and reflection are vital inasmuch as the variety of interpretations and ideas can enrich the source of solutions and other research questions. Finally, the topic raised should be deepened by further constellations around the addressed issue; a multi-stakeholder integration is preferable yet not necessary to anchor knowledge. This is in line with UNESCO's (2014: 126) understanding of a whole-institution

Figure 7.1 Transdisciplinary process in teaching and research

approach: "Collaboration and partnerships between university researchers and community stakeholders should be scaled up as mechanisms to deepen learning, strengthen the knowledge base on local social, environmental and economic issues and contribute to solutions for local-level sustainability."

Thomas (2016) stresses that organisational members engaging with education for sustainable development often have limited influence on higher-level policies. In the event that a whole-institution approach is not supported by an institution or is faced with scepticism, systemic structural constellations are also useful to generate bottom-up change. Integrating structural constellations addressing sustainability issues within teaching and learning contexts, by offering learning opportunities, has an impact on the curriculum as well as on students and their competencies. Focusing the sustainability impacts of the campus and its operations in learning and researching contexts, the students might also influence changes in institutional operations. Moreover, by analysing multilevel, complex sustainability challenges, learners will be enabled to recognise dysfunctional boundaries as well as cut value streams, so that they can manage sustainability issues in a more clear-sighted, comprehensive, cross-cultural and interdisciplinary way (see Table 7.2).

Table 7.2 Didactical opportunities of systemic constellations

Learning issues	Sustainability requirements	Whole-institution approach
Integration of different types of learning (as by doing, by reflecting, experimental, problem-based, collaborative, etc.)	Enables depiction of complex and systemic issues	Holistic, transformative and inter-/ multidisciplinary
Integrative and participative science or interaction	Inter-, multi- and transdisciplinary topics can be addressed	Values, reflection, critical thinking and problem solving
Applying different types of knowledge	Competence enhancement possible	Participatory reflection, discourse and decision making
Formal as well as informal learning	Focus on different synthesis topics (Rickinson and Reid, 2016)	Locally relevant and applicability in daily life settings

Thus, the constellations' outcome can be closely connected to the so-called science shop.[1] The members of the platform *livingknowledge.org* describe it as small entities carrying out "scientific research in a wide range of disciplines – usually free of charge and – on behalf of citizens and local civil society". Therefore, constellations can be used as an applicable method for inter-, multi- and transdisciplinary research and transfer that are society driven. In this sense, transdisciplinary settings are not necessarily integrated into teaching or research contexts, but practitioners or society get space for the participatory investigation of social concerns in a scientific surrounding.

Note

1 Living Knowledge, The International Science Shop Network, www.living knowledge.org/science-shops/about-science-shops/.

References

Barth, M., 2015. *Implementing Sustainability in Higher Education: Learning in an Age of Transformation.* Routledge: New York.

Evans, N.S., 2016. 'Implementing education for sustainability in higher education through student-centred pedagogies'. In: M. Barth, G. Michelsen, I. Thomas, and M. Rieckmann (eds.), *Routledge Handbook of Higher Education for Sustainable Development.* Routledge: Oxford, 445–461.

Hirsch, G.H., Hoffmann-Riem, H., Biber-Klemm, S., Grossenbacher-Mansuy, W., Joye, D., Pohl, C., Wiesmann, U., and Zemp, E. (eds.) 2008. *Handbook of Transdisciplinary Research.* Springer: Heidelberg.

Parker, J., James, P. and Atkinson, H., 2004. 'Citizenship and community from local to global: implications for higher education of a global citizenship approach'. In: J. Blewitt and C. Cullingford (eds.), *The Sustainability Curriculum: The Challenge for Higher Education.* Earthscan: London, 63–77.

Thomas, I., 2016. 'Challenges for implementation of education for sustainable development in higher education institutions'. In: M. Barth, G. Michelsen, I. Thomas, and M. Rieckmann (eds.), *Routledge Handbook of Higher Education for Sustainable Development.* Routledge: Oxford, 40–55.

UNESCO, 2014. *Shaping the Future We Want.* UN Decade of Education for Sustainable Development (2005–2014) Final Report, Paris, France.

8 Taking up the final cudgels for systemic structural constellations

In systemic structural constellations, systems can be simulated by spatial arrangements of persons or symbols. The success of the method lies in assigning the action research and can be described in terms of the systematic spatial locations and perceptions of decision-makers (Varga von Kibéd/Sparrer 2014; Schlötter, 2005; Sparrer/Varga von Kibéd, 2001). System constellations permit both a deep look into the informal structures and relationships of institutions and social structures as well as the testing of interventions or different solution options with regard to their effects. Systemic constellations can be used specifically in the context of academia, demonstrating effective starting points to overcoming resistance and to addressing specific academic content interdisciplinarily, reflexively and experientially. In addition, content- and structure-related recommendations concerning the implementation of sustainability in academic workaday life can be developed. The gain in constellation work can be seen in systemic perspectives illustrating the context overlaps or interferences of several logical progressions, descriptions and attributions, as well as circular interdependencies. In teaching contexts, it engages "students in a transformational process by encouraging critical reflection on their learning and actions" (D'Andrea and Gosling, 2005).

In fostering sustainability, it should be always clear that sustainability is one topic among others (like quality, development, finance, etc.) and has to be integrated on the basis of systemic principles within an institution. In complex contexts, like sustainability and resilience, learning progress can be achieved on the knowledge level as well as on the level of action. So by integrating the tool, structural constellations can offer a rapid problem identification and problem structuring in order to allow direct problem solving in cooperation with administration and science. The method appears to be a very powerful tool in explaining and transferring multicausalities in systems and offers itself as a complement to traditional methods. Findings, implementation options, conclusions and other results from work with systemic constellations are often not attainable by a pure study of documents, interviews or an empirical survey, at least not at a comparable speed. The visualisation of depth structures and hidden patterns allows a discussion of ways in which academia might transition towards more sustainability. Its use requires sound training in systemic structural constellations and a respectful interaction with people and their concerns. Inasmuch as the method has the potential to build a bridge between administration, science, innovation and change for identifying and initiating new options, on the one hand, and ways for action and implementation towards sustainability in academia, on the other, an ethical and respectful treatment of concerns, people and processes is mandatory.

References

D'Andrea, V. and Gosling, D., 2005. *Improving Teaching and Learning in Higher Education*. Open University Press: Maidenhead, UK & New York.

Schlötter, P., 2005. *Vertraute Sprache und ihre Entdeckung: Systemaufstellungen sind kein Zufallsprodukt – der empirische Nachweis*. Carl-Auer-Verlag: Heidelberg.

Sparrer, I. and Varga von Kibéd, M., 2001. 'Systemische Strukturaufstellungen: Simulation von Systemen'. *Lernende Organisation* 4, November/December, 6–14.

Varga von Kibéd, M. and Sparrer, I., 2014. *Ganz im Gegenteil. Tetralemmaarbeit und andere Grundformen Systemischer Strukturaufstellungen – für Querdenker und solche, die es werden wollen*. Carl-Auer-Verlag: Heidelberg.

Index